Eruptive Fevers

A History of Medicine - Scarlet Fever, Measles, Small-Pox and Treatments in the 19th Century

Being a Course of Lectures On

By William Vallancey Drury

Logo art adapted from work by Bernard Gagnon

ISBN-13: 978-0-359-73314-9

First published in 1877

Contents

Preface v
History and Characters of Eruptive Fevers 7
Scarlet Fever 13
Measles 39
Roseola 48
Rötheln 49
Chicken-Pox 51
Erysipelas. – (*St. Anthony's Fire*) 52
Small-Pox 57
Inoculation 78
Vaccination 82

Disclaimer

The contents of this book represent medical diagnoses, procedures and treatments of the nineteenth century. The publisher does not condone, advise or assume any responsibility for this book's use in any medical context.

The contents herein are presented purely for informational and historical purposes.

Preface

With a view to making these "Lectures on Eruptive Fevers" more generally useful to students of Homoeopathy, and to medical men who might wish for a book of reference on the subject, I have prepared them for publication.

They were originally delivered in a short course at the London Homoeopathic Hospital, Great Ormond Street. They have since then received the addition of much fresh matter.

I have gladly availed myself of the observations of others where such seemed reliable, and have specially to acknowledge my obligations to the works of Dr. Aitken, and of my friend the late Dr. George Gregory, whose large experience as Physician to the Small Pox Hospital for many years made him one of the leading authorities in this country.

His book on Eruptive Fevers, though now a little out of date, contains such a large amount of valuable information, that no subsequent work on the same subject would be complete without reference to it, though I have had occasion to dissent from some of his opinions.

The division into separate lectures has not been retained, though I have adhered to the mode of expression used when they were given, feeling that as they were intended for teaching, such language was most suited for the student, and will not be unacceptable to the practitioner, who may be glad sometimes to have a decided expression of opinion as to why one medicine should be used in preference to another.

I have endeavoured to give what I hope may be considered, a correct view of what homoeopathic practice should be; and while claiming perfect freedom to cast off anything that had the appearance of bondage to a system, have at the same time tried to show that a pretty close adherence to homoeopathy was more likely to give better results than any careless mode of practice that allowed a resort to all sorts of measures at the whim of the practitioners.

The much-vexed question of dose I have hardly touched on, only on one or two occasions naming the strength I had used or recommended under special circumstances.

After many years' experience, the rule I have adopted is this, — that while trying to secure the greatest amount of good from the dilution of medicine, the nearer this can be obtained to tangible doses the better;

and if a small tangible dose of say from 1x to 3x or 3 will give equally good results, such strength is preferable in a higher dilution. At the same time, where experience is in favour of the higher, such should be used.

I have thought it desirable to go at some length into the subjects of inoculation and vaccination, as, owing to the last and present epidemics of small-pox, these questions have assumed increased interest. The opinions expressed may vary from those of many, but if they in any way help to the formation of more definite views as to what cow-pox really is, a great object will be gained.

Much that I have said is not new, but rests on facts that if acted on will lead to our obtaining an unfailing supply of reliable vaccine.

7, Harley Street, Cavendish Square
April, 1877.

History and Characters of Eruptive Fevers

Gentlemen — In the short course of lectures on exanthematous diseases that it falls to my lot to deliver, it is not my intention to include all that have been so classified or that might fairly come under that category, but rather to confine myself to those eruptive fevers that are commonly recognised as such.

The name of these diseases is derived from the Greek, and signifies to bud forth, to effloresce, as from the budding forth of a flower.

The diseases in question, being characterized by the breaking forth of an eruption, are known as the exanthemata. From this derivation you will see at once what a wide and varied group of diseases might be classed under this heading. Bateman adds urticaria, purpura, and erythema to the list. Even the plague has been so classified, but by common, though unexpressed, consent the number has been restricted, and I propose, as I have already intimated, to avail myself of this arbitrary but wholesome rule, and confine my remarks to scarlet fever, measles, roseola, rötheln, chicken-pock, small-pox, and vaccination.

Before proceeding to speak of any one of these, it may be well to take a glance at the group generally, and see what features they possess in common, in their general character, as well as in their history.

Now eruptive fevers, as their name implies, are characterized by the presence of fever and the appearance of an eruption. The fever precedes the eruption, and in some there is abatement on its appearance.

There are symptoms, such as the catarrh of measles, the sore throat of scarlet fever, and the backache of small-pox, that may give us some idea what form of an eruption the premonitory fever is to usher in, though the fever itself may not tell much, as the initiatory stage of all these diseases have many features in common, whatever the budding forth of the efflorescence may turn out to be.

Besides the help that we may get from special symptoms such as those alluded to, there is generally something about the look of the patient very difficult to describe, but which conveys some information to the experienced eye as to what may be coming. In addition to this, if any epidemic is prevailing at the time, or if the patient has been exposed to any particular infection, our suspicions will naturally point in that direction, though it will not be wise to jump to any hasty conclusion without other corroborative evidence to support this.

Parents are always anxious to know early what the nature of the illness is, not merely from a natural anxiety to know what they have to fear, but also that they may know what precautions it will be necessary to take to guard against the further spread of illness, On this account, once you have satisfied your own minds as to what you have to deal with, it is well to impart your information to the friends; but if from the undeveloped state of the case you can only form a surmise, it is well even then, if your suspicions are roused, to give a little note of warning that may insure some precautionary measures being taken; if the threatenings blow over a little, extra caution will have done no harm; if, on the other hand, your conjectures are right, the friends will be all the better pleased that some precautionary measures have been taken in advance.

It is very possible that now and again some foolish and inconsiderate person may resent as a wrong being told that a danger threatens if it does not appear afterwards, and no doubt this unreasoning spirit may cause you a little worry; but if you have exercised the judgment God has given you to the best of your ability, rest satisfied that you have acted rightly. If in the one case you are wrongly growled at — for I think that word expresses the hints and remarks that will be made for your benefit — in other similar cases you will be commended.

Care, however, is needed not to raise an alarm rashly. It is always well to be sure of one's ground in expressing an opinion; and if this is needed in speaking of what is coming, it is far more necessary in pronouncing on what is. It is awkward to pronounce a rash to be measles to-day and something different to-morrow. These mistakes are, however, made. I have no doubt many a case has been sent to an hospital said to be one form of complaint that turned out to be something quite different. When we hear of such a mistake being made, we should be lenient in our judgment, remembering some difficulties that we, like other medical men, may have experienced in our own practice.

Let us hear the evidence of a man who could afford to admit a difficulty of this kind. The late Dr. George Gregory, speaking of roseola, thus alludes to cases where scarlatina and variola seem each to threaten:

"The occurrence is very annoying in practice. The physician first pronounces that his patient has fever. Two days after he changes his note, and informs the friends the patient, besides fever, has the roseola or rose-rash — an affair of no consequence. Two days after that he announces to the astonished listeners that his patient has small-pox! This," says the doctor, "once occurred to myself in consultation with Mr. Hammond, of Windsor. Three diseases in as many days!"

To return, however, to the points of resemblance in exanthematous diseases. They are diseases of childhood. The cause of this is very evident; they are infectious, and often epidemic, and those who have gone through one attack rarely suffer from a second. The young also seem to be more susceptible than those that are older. Children, therefore, when exposed to infection are likely to suffer; adults, who have probably gone through the disease in childhood, escape. Owing to vaccination, small-pox is now rare among children; but for this protection the rule would be found to hold good, and large numbers of children would suffer.

Going back to a time when vaccination was less rigorously carried out than at present, "there died in London of small-pox, during the two years 1840-41, 2,286 persons, of whom 2,060 were under, and only 226 above, 15 years of age" (Gregory, on "Eruptive Fevers," 87).

All these diseases have a period of latency, so far as outward appearances go, before they break forth into activity, like the seed under ground that swells, and bursts, and sprouts, so eruptive fevers, once the infection is sown, pervade the whole system till, in obedience to the law that regulates their course, the time comes for their efflorescence. This characteristic, of growing within the body, like leaven, leavening the lump, has caused them to be classed in the death lists of the metropolis as zymotic diseases, from the Greek word signifying "leaven."

During the latent period, or at least some portion of it, they have the power' of extending their influence to those who come within range of their operation. In this way these diseases spread, and we see the reason why in schools the children fall sick in; detachments. Hence the necessity of isolating those who show the slightest signs of illness of a suspicious character.

A young gentleman was taken ill with measles, his sister was immediately separated, but was allowed to continue her attendance at a French class. Her mother was in the habit of accompanying her, but as she had to attend to her son she was unwilling to go to the class for fear of spreading infection. A message was sent to me telling me of this caution on the part of the mother. I sent her word that she might go to the class when she pleased, but that her daughter had better keep away, and that she probably would have the measles on a day I specified. This prediction proved correct, and my fears about the daughter's spreading infection were verified by others taking it.

It is, therefore, a wise precaution at once to separate those who have been exposed to infection, till the longest known period of latency has passed.

As it will be necessary to refer again to this subject when speaking of the different eruptive fevers, I need not dwell longer on it at present.

Another feature common to some of these diseases is their tendency to leave some mischief behind. The sequelae, or *dregs,* of popular language, are often exceedingly troublesome. In this way we may have distinct mischief, such as dropsy following dose on scarlatina, forming, as it were, part of the illness; or we may have broken health referable to an attack of measles that occurred months before Vaccination gets no small share of abuse for what it is accused of doing.

When we are aware of the dangers patients run from these after-consequences, it is the duty of the medical man to watch his patient through this period— not so closely as to give rise to any possibility of his motives being misinterpreted — but as he knows the danger, which his patient or the friends may not, he must not, through over-sensitiveness on. his part, allow his patient to run any risk, or allow himself to be charged with negligence for not watching and foreseeing the advance of what may prove to be serious mischief.

The next point of similarity is the tendency exanthematous diseases have to appear as epidemics. Isolated cases of greater or less severity may be appearing all the year in large towns, or in a given area of country, but from time to time large numbers are stricken down together, and the exanthem, whatever it may be that has thus assumed the epidemic form, becomes the prevailing disease, and if severe may largely increase the death-rate of the district for the year.

Even the history of exanthematous diseases furnishes a point of resemblance. They are not unfrequently said to have made their first appearance in comparatively recent times. Thus the late Dr. George Gregory says: "All the exanthemata have sprung up since the commencement of the sixth century," he, along with others, assuming that, because we have no accurate description or authentic records, these diseases had not been seen before, or because in the history of some epidemic, as related by non-medical historians, we fail to identify the oases spoken of with some disease with which we are now familiar, they must be different — to my mind a very unwarrantable conclusion.

On the other hand. Dr. Richardson, from the description that has been handed down to us of the plague of Athens, as told by Thucydides, has arrived at the conclusion that this was a terrible visitation of malignant scarlet fever. Dr. Francis Adams, the editor of the works of Hippocrates, takes a different view. He thinks that the pestilence which prevailed during the Peloponnesian war partook of the nature of glandular plague, reminding us at the same time, as a difficulty in the way of this opinion, that Thucydides does not mention the presence of buboes, but that he was not a medical man, and that his description was of a general character. I cannot but think that this omission is nevertheless fatal to Dr. Adams's belief.

This scourge has also been supposed to have been epidemic small-pox — about as unlikely as to have been the plague. These conflicting opinions serve to show us the obscurity that hangs about the early history of the exanthemata, and the folly of fixing their first appearance in any given year.

I think Dr. Richardson takes the correct view that diseases maintain the integrity of their type, and believe that, if the distinctions of diseases were as well understood formerly as now, there would be no difficulty in knowing what each epidemic plague was.

In some cases the descriptions of the old historians are sufficiently accurate to enable us to arrive at a tolerably correct conclusion. Thus we may say that the Black Death of Edward the Third's reign was malignant typhus, and not object to the opinion of others who recognise cerebro-spinal meningitis in the writings of Procopius, who flourished about A.D. 500, in the time of Belisarius and Justinian.

When we remember that great medical writers come few and far between, and that even these great men may not have had the opportunity of seeing, some of the diseases with which we are familiar, and when we know how they confounded some of them together that appear to us perfectly distinct, we can readily see why it is that history fails us.

We must come to recent times to have diseases distinctly defined. Sydenham was the first to separate measles from small-pox, but twenty years later than his time Morton considers that measles and scarlatina are the products of the same miasm. As late as 1779 Withering speaks of measles as being nearly allied to scarlatina.

The varieties of scarlatina were not recognised as forming one and the same disease as late as Cullen's "Nosology," published in 1792; but the following year, 1793, Withering pointed out the identity of scarlatina anginosa and scarlatina gangrenosa.

It is within our own memory that typhus and enteric fever have been separated, and even yet there may be some points to be cleared up as to the fever spoken of as continued fever. And some may still think there is a connection between croup and diphtheria, though I dare say the majority of medical men will be of opinion that they are well defined and separate diseases, an opinion in which I concur.

With all this uncertainty on some points about which we might have expected less difficulty, we cannot but be struck by the clearness of description that characterizes some of the old writers. Some of the cases of fever described by Hippocrates are singularly graphic. The sweats, the enlarged spleen, the mortality among puerperal women, the crisis, and other features are more or less familiar to us, and we can readily believe that in parts of Greece we might find similar cases at the present day.

On the other hand, when he speaks of many who had their mouths affected with aphthous ulcerations; defluxions about the genital parts, ulcerations, boils externally and internally about the groins, watery ophthalmics of a chronic character with pains, fungus excrescence of the eyelids externally and internally, called fici, which destroyed the sight of many persons; fungus growths, ulcers, carbuncles, "other affections which are called the putrefactions," also "large ecthymata" and "large tetters," there are doubts as to what diseases are intended, and I think it probable that some exanthematous diseases are spoken of, the identity of which this famous physician had failed to recognise.

As most medical men have seen very severe as well as very mild cases of the same disease, so epidemics have varied greatly at different periods. At one time a disease is so mild as to create no anxiety, at another it has assumed a form of intense malignity. This has been especially the case with scarlatina, of which disease Sydenham gives but a brief account; he had not seen it in its severe form. Other writers, however, have told the rapidity with which death has taken place, of many children of a family being swept off, and of the severity of the symptoms as shown by the putridity and the character of the discharges, [1]

Measles as seen in our day is more dreaded on account of the possibility of chest complications than from its character as an eruptive fever; but measles has assumed a more severe form, as the recent epidemic in the Fiji Islands reminds us. [2]

It is, however, generally true of the exanthemata that they have usually visited the aborigines of different countries with great severity. This varying character must have operated with our predecessors, and may have frequently prevented their recognising a disease that at one time of its appearance had shown itself with a suppression of some of its chief symptoms, and at another with those symptoms developed in their fullest intensity, and they would naturally maintain that such diseases must be separate and distinct, [3] until long and careful study, as in the case of Withering (who devoted fifteen years to the inquiry), enabled them to arrive at a right conclusion.

Many speak of cholera and diphtheria as comparatively modern. I believe this arises from the want of an historian in former days. The account of the cholera as given by Sydenham reads wonderfully like what we know as cholera at the present day. The addition or intensification of a few symptoms would make the cases similar to Asiatic cholera, so that it requires no great stretch of the imagination to believe that the severer form may not have been a stranger to this island till 1831. The aphthous mouth of Hippocrates and some of the sore throats of past days in all likelihood had a diphtheritic character.

I think we may safely conclude that all our modern eruptive fevers, as well as other fevers, are but reproductions of what have been before, that they have appeared at different times in varying degrees of severity, and that the date of their first appearance is unknown. It may be that some of them arose during the time of the wanderings of the children of Israel in the wilderness, or at some other period in their history, when their sins were followed by the visitation of a "plague."

Whatever the immediate cause that made the words "the plague has begun" history, these diseases followed as a natural sequence, to that act of which Milton writes, when he tells,

> "Of man's first disobedience, and the fruit
> Of that forbidden tree, whose mortal taste
> Brought death into the world, and all our woe."

[1] A medical man whom I met in consultation in Boulogne a few years ago told me that they never had severe scarlatina there. All the cases were mild. I have no doubt he spoke only of what came under his own observation during a period of a few years.
[2] Even this week (April 26, 1876) we read of 59 deaths from measles in London — a very large death-rate for a disease that we speak of as much less fatal than scarlatina.
[3] Bateman considers the second edition of Withering's book in 1793 as the starting-point for the recognition of the different forms of scarlet fever.

Scarlet Fever

The disease of which we are about to speak very fairly derives its name from the appearance of the eruption that usually shows itself, though not always, for in some cases this characteristic symptom may be absent.

Another marked symptom that helps to give a name to one form of the disease is sore throat; like the eruption it is usually present, but it also may be absent.

The presence of gangrenous symptoms which appear in some cases also furnish a name for one form of this fever.

The name Scarlatina is frequently used instead of scarlet fever; it is not very clear who first used it. It is derived from "scarlatta," a red-coloured cloth (*Elliotson*), Its employment has led to a popular error that is not easily dissipated; thus we will be told that the complaint under which some one is suffering is not Scarlet Fever, it is only Scarlatina. Where the opportunity occurs, this should be corrected.

Scarlet fever has been divided into three species.

1. *Scarlatina simplex,* a mild form of complaint.

2. *Scarlatina anginosa,* a more severe form, where the throat symptoms are prominent on account of their severity.

3. *Scarlatina maligna,* which is marked by the presence of sloughing and great depression of the vital powers. A fourth kind is included by some — *Scarlatina sine eruptione.* This last division is misleading from the fact of the absence of the eruption being sometimes due to exactly opposite causes, in the one case the disease is so mild that, but for some accidental circumstance proving its existence, it might have passed unnoticed, in the other the patient is struck down at once and life extinguished almost before there is time for the development of the eruption.

There is no disease so full of anxious surprises, or more likely to be conveyed by a third party, or to be developed by some hidden but active source of mischief.

Scarlatina Simplex. — After a period of latency, the outward manifestation of the disease commences usually by a febrile attack. I say usually, for it may happen that the first intimation of illness is the nurse coming to announce that a child has a rash on it, or it may be that sore throat is first complained of.

Feverish symptoms, however, generally precede the affection of the skin. A child is observed to be less disposed to play about than usual, there is shivering or chilliness, headache, weariness, pain in the back and limbs, thirst, food is taken sparingly or turned away from. The second day, or it may be a day later, the rash appears, and with it the throat is complained of. On examining the face, neck, chest, and arms, there is a general look of redness, which on closer inspection is observed to be made up of a number of minute puncta or spots, giving the appearance of a boiled lobster. Some of the spots appear to coalesce; where it was appearing in patches it becomes more generally diffused, it spreads downwards, and in twenty-four hours the back, abdomen and legs may be covered with eruption. If it is not coming well out, the bends of the joints and the loins are good places to look for it. The redness disappears on pressure. On passing the finger over the skin it is often smooth, but in some cases a number of minute points, giving a sensation of roughness, may be felt, like the *cutis anserina,* or goose skin. This differs from the feeling of elevated patches observed in measles.

The eruption is usually at its height the fourth day, on the fifth it is declining, and may be gone by the seventh. MM. Barthez and Rilliet mention six to eight days as the ordinary duration of the eruption of scarlet fever in its normal form, that it requires a longer period to develop itself than that of measles, and is more persistent at its maximum, lasting twenty-four, and even forty-eight hours. The eruption may only last five days, or as long as ten, but never longer. (Gregory, American edition.) In some cases there is no eruption, only slight sore throat. I shall allude to them when speaking of *Scarlatina sine eruptione.* As the eruption is coming out it may suddenly recede and again come forward. On its disappearance the cuticle begins to peel ofif in the form of a scurfy desquamation; in some cases entire pieces from the palms of hands or soles of feet come away. The nails have been thrown off. There is often itching and tenderness. The desquamation may go on to the end of the second week, or be delayed longer according to the eruption. It may continue for about three weeks.

The tongue is at first covered with a white fur, but this gradually clears off, when it is observed that the tongue is red, the papillae elevated, giving it a strawberry appearance; about the fauces, and above the uvula there are some red spots.

The fever shows but little abatement on the coming out of the eruption, and remains as long as it does, or longer, if the throat continues sore. The pulse is usually very quick; even in mild cases it may be over 130. The temperature may be from 102 to 105 or 106; it has even been noticed as high as 112. The thermometer affords such valuable information as to the progress of fevers, often giving an early note of warning where danger threatens — that those who from careful observation have learned to appreciate it, keep a careful register of its markings and reap the benefit of so doing; while those who undervalue it suffer a corresponding loss. There is usually an increase of fever in the evening, and, owing to this, there is not unfrequently some wandering at night. If the case is a mild one, the rash will disappear, the head symptoms pass away, the temperature fall to the natural standard, and the soreness of the throat gradually cease, without the development of any of those urgent symptoms of which we have yet to speak, and which characterize the more severe forms.

Some of the symptoms of this disease are of very different significance; thus one symptom, a miliary eruption, may show itself without adding to the severity of the disease, while another, that requires careful watching, is enlargement of the glands of the neck; this may accompany scarlatina simplex, and pass away without giving trouble, but as it is a symptom of considerable importance in the more severe forms of the disease, and as it may be a source of mischief at a later period, even after mild fever, under no circumstances should their presence be overlooked.

The danger incurred by pregnant women when attacked with scarlet fever is great, as also by women in childbed, and the danger of conveying this disease ought to make medical men who engage in general practice extremely cautious in approaching the lying-in room if they have been in attendance on scarlet fever within any short period.

It must have been a mild form of scarlet fever that made Sydenham speak of what he saw as " the name of a disease; " still he gives a warning about a more severe invasion of it, and against its being treated "too learnedly" a caution that has not always met with the attention that it ought.

Scarlatina Anginosa, — Where the disease assumes a more severe form as shown by the manner in which it attacks the throat, it gets the name of *Scarlatina anginosa.* There is now considerable swelling of soft palate, uvula, and tonsils, the redness is intensified, there is stiffness of the neck, the submaxillary and parotid glands are enlarged and painful. The glandular swelling may precede the sore throat, or may not appear till the throat is healing. Pressure, in some cases even touching an enlarged gland, causes pain. The throat is seen with difficulty. In the case of children there may be a painful struggle to secure a proper inspection. If the child is wrapped up in a blanket, the arms being secured by it, and then allowed to sit on the nurse or mother's knee, the head being steadied by the nurse's hand, or some one else's, the mouth can be opened with the spoon and the examination made. At night the doctor may have to hold the spoon with one hand, and the candlestick with the other. All this may seem needless to explain, but when we see how some men are perplexed in dealing with their patients, and, if it be a child, witness the protracted struggle that takes place, the fighting and screaming of the little sufferer and the agony of the mother, any directions that tend to prevent this are not out of place. I learned the use of the blanket many years ago from the late Dr. Jacob, of Dublin, who in operating on children's eyes, before the days of chloroform, rolled his little patient up like a mummy, and then performed his operation. Whenever you have anything of this kind to do, make up your mind beforehand how to carry it out have all your appliances ready, and then firmly and quickly, but with all kindness and sympathy, do what is necessary.

As the disease advances the symptoms will all be aggravated; there will generally be more headache, and more delirium. The delirium is sometimes very violent. Dr. Schneider, of Magdeburg (*British Journal of Homeopathy*, vol. xiv.), in Herschel's Zeitschrift vol. v., No. 5, makes some interesting remarks on delirium. He says: "The typhus blood acts similarly to blood charged with alcohol, not only on the cerebrospinal, but also on the ganglionic nervous system. In inflammation the brain may suffer in two different ways; either by an extension of the inflammatory irritation to its substance, or by pressure of blood. Sleeplessness and delirium indicate the first, sopor and insensibility the second case, which is similar to the effect of water in the cerebral cavities, and apoplectic extravasation. A patient in the soporous state, induced by inflammation of the brain, scolds in his sleep, and occasionally wakes up of himself, or when roused, in a delirious state. On the contrary the hydrocephalic lies quiet, as in natural sleep, and the difficulty of rousing him goes on increasing as he lies (at a later period of the disease), completely devoid of consciousness, with open (often squinting) eyes, and in a convulsed state. The apoplectic again, from the commencement can either not be roused at all or only momentarily, to a half state of consciousness. Children. with very irritable brains, become delirious even in catarrhal fever, and adults otherwise not disposed to delirium in the hot stage of intermittent fever. No other poison produces these various acute diseased states of the brain in a more marked manner than the scarlatina poison. In scarlatina the cerebral irritation bears a relation to the fever, increasing and declining with it. Of all exanthematous diseases scarlatina is most frequently accompanied with delirium."

I have given Dr. Schneider's remarks at some length as they are very interesting. We expect violent delirium or a good deal of raving with the high fever of the anginose variety, but in the malignant variety, after some excitement, we look more for oppression of the brain, from the overpowering influence of the poison. The temperature maybe over 106, perhaps 108. Swallowing is very difficult; there is hoarseness or alteration of voice. The eruption may be suppressed, or of a darker, duskier hue, or it may recede. Not unfrequently it is a day later in appearing. Cellular inflammation may set up about the ear. Respiration may be impeded from the mischief in the throat. An acrid ichorous discharge may flow from the nostrils, a purulent discharge may flow from the ear; sloughing may take place, followed by deafness; children are often brought to us suffering from deafness due to an attack of scarlet fever some months or years previously.

The mischief in the throat continues, there are deep foul ulcers, sloughing. Owing to the increase of cellular inflammation and swelling, it becomes almost impossible to see the throat, and the state of it can only be guessed at. The mouth is sore; the attempt to introduce the spoon causes bleeding. There are sordes on the teeth, and the fever continues; this it will do, though the eruption has gone. At first it was associated with the state of the skin, now with the throat; and it will continue till the sloughs separate, which may not be till the fifteenth day. The high fever and the sore throat are the marked symptoms we have to deal with. Dr. H. Goullon (*British Journal of Homoeopathy*, vol. xxiii., p. 64) speaks of a case of glandular swelling of the neck where an abscess opened and discharged its contents. He says the destruction of the soft parts, especially of the cellular tissue, was so great that there was the appearance of an anatomical preparation of the superficial muscles. The larynx was exposed to view. The left parotid, which had been swollen, now increased, attended with great destruction of the soft parts (otorrhoea also). As a consequence of the destruction, a large blood-vessel was injured, and the child died from haemorrhage in the fifth week of her illness.

In connection with this condition of things a question arises, may we have diphtheria present at the same time? I think there is no doubt about it. A diphtheritic exudation will be seen from time to time. As some medical men would dispute this, I shall call two or three independent witnesses, who wrote, no doubt, after diphtheria was described by Bretonneau, but before it had made its appearance again as it has done of late years. In an admirable work on diseases of children, by the late Drs. Maunsell and Evanson, in the third edition, published in 1840, we read: "Upon looking into the throat we may find considerable redness and swelling, and often portions of lymph thrown out upon the tonsils and uvula, which may at first be mistaken for ulcers or sloughs of the mucous membrane, but are not so, as upon their removal by gargling the latter will be perceived to be whole and unbroken" (page 391). Another witness, the late Professor Alison, of Edinburgh, says: "Before the proper symptoms of the scarlet fever have abated, various effects may result from the affection of the fauces, by which the danger of the disease may be much aggravated; the inflammation may extend down the larynx, lead to the effusion of a flocculent false membrane, and cause the symptoms of croup " (Alison's "Pathology," 484). Other writers make similar statements. Dr. Gregory says, " Should the fever be active, portions of coagulated lymph will be seen effused. These are often mistaken for ulcers; but in many most severe cases of anginose scarlatina there is no actual breach of surface — only excessive engorgement, with effusion of lymph" (*Eruptive Fevers,* 156). Dr. Goullon, speaking of the epidemics he witnessed at Bemda, says some of the cases assumed a diphtheritic character, adding to their gravity. Dr. Graves says, "Croup occurred in two instances in which, notwithstanding the opinions of M. Trousseau, I could not doubt its origin in scarlatina. It happened, no doubt, in cases which had exhibited the diphtheritic patches, without much surrounding inflammation on the tonsils, but the eruption was sufficiently marked to remove all obscurity. One child, who recovered, ejected the false membrane (which I still preserve) in a tubular form, and presenting a cast of the trachea a little beyond its bifurcation. In the child before mentioned who died, patches of false membrane were also ejected; but she sank exhausted, and the disease was afterwards discovered to have extended far into the bronchial ramifications " (Clin. Med., vol. i., 321). Dr. Graves speaks of these cases as croup, but, whether they were so or not, they are of much interest, and it is of course a question whether they were not of a diphtheritic character. The late Dr. MacLimont relates a case in the *British Journal of Homoeopathy,* vol. xxi, 1863, of a boy ill of scarlet fever where there was diphtheritic deposit. He applied chlorate of potash, and used the first centesimal dilution of protiodide of mercury: the boy did well. As a matter of scientific inquiry, as well as for its practical value, it is well that the possibility of such complications should be known, but as homoeopathic practitioners, select their medicines in accordance with the whole group of symptoms present in each case, increased knowledge of the disease ought to bring with it increased accuracy in treatment, though unhappily here the new symptoms are attended with increased danger.

Scarlatina Maligna, — Angina Maligna Putrida, Cynanche Maligna. "The sore throat attended with ulcers" of Fothergill, who thus described the epidemic of 1747-8, is one of the most formidable complaints that it falls to the lot of the physician to encounter. Some medical men have seen many cases, others none. In 1846 Dr. Elliotson says, "So comparatively rare, however, is this species of the affection, that I have never had occasion to treat a patient with it." He could say at that time that he had lost but two cases from scarlet fever, and had heard other practitioners say they had never lost a case from it" ("Pract. of Med.," 417). Since Dr. Elliotson wrote this we have had ample opportunities of seeing malignant scarlatina in London, and so, no doubt, Dr. Elliotson had before he retired from practice. Dr. Graves, speaking of an epidemic that in 1801 — 1804 committed great ravages in Dublin, says, from information he obtained from Dr. Percival, death sometimes took place as early as the second day: after that period for twenty-seven years it was of a mild form, so much so that out of eighty boys attacked in a public institution not one died. The great point of interest connected with Dr. Graves' narrative is this, and it should not be lost sight of: those who saw scarlatina during "the years of grace" came to the conclusion that the diminished mortality was entirely attributable to the cooling regimen, and to the timely use of lancet and aperients which had been interdicted before, not a few of the earlier men being disciples of the Brunonian school Dr. Graves says, "This I myself learned, this I taught, — how erroneously will appear from the sequel." "It was argued that had the cases which proved fatal in 1801-2 been treated by copious depletion in their very commencement, the fatal debility would never have set in, for we all regarded this debility as a mere consequence of previous excessive* reaction. The experience derived from the present epidemic has completely refuted this reasoning, and has proved that in spite of our boasted improvements we have not been more successful in 1834-5 than our predecessors in 1801-2."

In the malignant form of the disease the tendency to slough is very marked. The first case of malignant scarlet fever that I recollect seeing somewhere about 1842 was that of a young girl in a poor dwelling house near Gillespie's Hospital in Edinburgh. The rapidity and the putridity impressed me strongly. A scratch of a pin on the arm and an abrasion on the forehead — how produced I cannot say — were the following day surrounded by comparatively large sloughs. Being then either a student or barely out of my student days, I asked a medical friend to see the case; he suggested an aperient containing turpentine, saying, however, that the child would probably die when the bowels acted. I preferred allowing that event to take place without the aid my friend suggested. The case is a good illustration of "everything being done," so satisfying to the lovers of big doses.

In malignant scarlet fever the rash is usually late in coming out, or is suppressed; it may disappear and return again; it has a dark livid colour; the face has a dark congested appearance with heaviness of the eyes; petechiae may be present. The throat has a dark red hue; the fauces are swollen, and over them and the tonsils there is a quantity of viscid mucus or purulent discharge; this may be sputtered out as the throat is looked at. In some cases the tongue is dry and parched, the viscid mucus forming sticky strings without imparting moisture. Ash-coloured slough appear on the tonsils and uvula; there is acrid discharge from the nose, and fetid breath, as well as generally unpleasant odour about the body. The countenance is dull and heavy, the voice hoarse and hollow, the respiration laboured and hurried. The acridity of the discharge from the nose causes soreness about the nostrils and mouth, and the fingers, if they come much in contact with it. The patient endeavours to cough up the sloughs or mucus, or will with his own hands endeavour to clear away the viscid mucus from the mouth. There is often extensive destruction in throat from sloughing, or as in the case already alluded to sloughs form rapidly on various parts. The mind becomes clouded, picking of the bedclothes and wandering are present; in some cases there is much delirium; if an answer is elicited, it is obtained with difficulty. Coma supervenes, in some cases convulsions, and often at an early stage death closes the scene, though in other cases life is prolonged for some days.

Scarlatina sine eruptione or latens. — This name is usually applied to those mild cases that often escape notice till some symptom, such as anasarca, leads to inquiry. I have seen a case pass through the disease with hardly a trace of illness, followed by symptoms of sudden effusion on the brain. I have also seen several members of a family ill with slight, sore throats and just a suspicion of scarlet fever, till at length one was attacked with sufficient severity to show what the real nature of the former cases had been.

I have recently attended a case of scarlet fever where the nurse who was in attendance was attacked with sore throat, followed by suppuration of both tonsils. There was no rash, nor fever to speak of; she was able to discharge her duty in attending the sick child. On a former occasion, when nursing a child with scarlet fever, she was attacked with sore throat Were these two attacks simply coincidences, or were they produced by infection? My belief is that they were caused by being in the neighbourhood of the fever. We see this sufficiently often to prevent our regarding them as accidental occurrences unconnected with the disease.

In whooping-cough we will at times find a paroxysmal cough attacking some one in attendance who has had the disease previously. I have been more inclined to regard this as a nervous copying of the cough, rather than as produced by any susceptibility to the disease. In the case of scarlet fever the sore throat is not a thing that can be copied, at least not when such an objective symptom as suppuration is present. In spasmodic whooping-cough it may be copied, as chorea, hysteria, and epilepsy are.

I saw a child, a few years ago, attacked with dropsical symptoms; this led me to make inquiries, when I found that symptoms that ought to have attracted attention had hardly been noticed, and but for the attack I speak of, and the inquiry I was led into, the parents would have been in happy ignorance of the dangerous visitor that they had in the house. Speaking before of this form of disease without eruption, I have included the severe cases that die before the development of eruption. In these cases the system gets, as it were, such a dose of the poison, that in almost a few hours it succumbs, crushed beneath a weight it cannot struggle under. It is however clear that there is no similarity whatever between the mild cases, where the eruption is absent from the very mildness itself of the attack, and those other cases where there is no strength to throw out the eruption; hence this last classification is of no practical value, beyond its use in showing those who are studying the disease, the varieties of symptoms, end cases they may meet.

Complications. — I have now gone through the ordinary forms of scarlet fever, and I imagine most medical men who have been any length of time in practice have met with every variety. Of course there may be many complications present, such as meningitis, erysipelas, pneumonia, convulsions, hepatitis or jaundice, diarrhoea, and other accompaniments that require careful attention in their treatment, and in no small degree increase the danger. A complication noticed by Dr. Barnes and Sir John Cormack is the occurrence of vaginitis, accompanied with muco-purulent discharge. This, perhaps, should more properly be included among the sequelae.

Sequelae. — Scarlet fever is a treacherous disease, and perhaps in nothing is this more marked than in the fact that where a case has been so mild as to be hardly recognizable, life may yet at a later period be seriously jeopardized or lost by the occurrence of dropsy. This often happens after the mildest cases. It may show itself in the fourth week. The patient will be observed to look pale, then a little puffiness will be noticed beneath the eyes, the skin of the face will have a waxy look, the feet and ankles swell. In severe cases the swelling extends to the whole body. Hydrothorax, hydropericardium, ascites, or hydrocephalus may be present. Distress of breathing comes with the first, while convulsions and coma show the presence of the last.

Where head symptoms supervene with a child that has shown no threatening of them before, if there is any suspicion of it, it is well to ascertain whether scarlet fever can have anything to do with their production. Do not leave the parents ignorant of the presence of great, and it may be immediate danger.

The urine is of course an important guide as to the approach of desquamative nephritis. If the presence of albumen is detected, or if the urine ' begins to diminish in quantity, danger of dropsy may be apprehended. If mischief progresses, smoky or reddish urine, passed in small quantity, in some cases with irritation, will be present. Casts of the tubuli uriniferi, and epithelial scales may be found in the urine. The smoky urine is due to the presence of blood and also, the late Dr. Atkin says, to the presence of numerous minute crystals of uric acid.

I have already spoken of glandular swellings, accompanied by suppuration and extension of the cellular inflammation to the ear; this spreading by the Eustachian tube may lead to the destruction of the bones of the ear, perforation of the membrana tympani, and permanent deafness, or to further extension of mischief to the brain and its membranes; death resulting from meningitis or purulent effusion.

In some cases swelling of the joints, as from an attack of acute rheumatism, supervenes.

Ulcers, bed sores, and abscesses, also affections of the internal organs, such as pneumonia, and other inflammations, are to be apprehended. Loss of sight has followed scarlet fever; this is a rare result.

I have more than once seen a secondary fever presenting the characters of continued fever, with typhoid symptoms, follow scarlet fever. Many years ago, not long after I had avowed my belief in the truth of homoeopathy, a gentleman came to ask me to see two of his children ill with scarlet fever, stating that a medical man who had been in attendance had told him the cases were severe. As he wished to have them treated homoeopathically, my advice was sought. This was to me a novel experience, being called in because the cases were severe, the allopathic attendant retiring. Having a painful recollection of what scarlet fever was during my nine years of allopathic practice, I entered more hopefully on my duty than I could have done if obliged to resort to the weapons that had failed me before. Both cases, that of a boy and a girl, were of very great severity; both were followed with fever such as I have spoken of. The little boy had erysipelas of the face, purulent discharge from the ear, and dropsy. I cannot now recollect whether the little girl had dropsy or not; but I well recollect for about six weeks looking each day as I approached the house to see if the window shutters were closed. Through God's goodness the children were spared. The pleasure of seeing my new remedies successful was great, but I doubt if any fee compensates the medical man for the anxiety such cases bring with them.

The only remaining trouble that I have to mention is diarrhoea, which may retard recovery by its weakening effects. Gaping may be present, and in severe cases ulceration of the bowels and bloody stools, and, of course, with such symptoms the danger is imminent.

The great debility that follows a severe case can hardly be looked upon as one of the sequelae; it is that debility, in connection with other mischief, that will cause alarm.

Appearances after death. — Fibrinous clots are found in the right side of the heart when death takes place early. This has led to the recommendation of ammonia in repeated doses in allopathic practice. Swelling of the solitary and agminate or clustered glands 'of Peyer, inflammation of the alimentary canal, kidneys, and brain, purulent collections. These will be looked for according to the symptoms observed during life. From what we may find in this disease I cannot but be struck with the resemblance it bears to the appearances I have seen in an inspection of the body of a cow that had died of the cattle plague which visited us some years ago. There the combination of pulmonary and intestinal mischief with a diphtheritic exudation, were the same as under certain conditions we might find in scarlet fever.

Diagnosis. — A well marked case of scarlet fever will not give much trouble; the character of the rash already described will prevent its being confounded with measles. But it is just in those cases where the rash does not come well out that difficulty is likely to occur. The high temperature, the sore throat, the evidently suppressed rash where its presence cannot be detected, must help us to form our opinion. Roseola, which resembles scarlet fever, is said to show the rash more on the chest. Rothëln, the hybrid of scarlet fever and measles, where the rash represents scarlet fever, but the catarrhal symptoms are those of measles, will be recognised by the mixed character of illness.

Propagation of the Disease. — There is no doubt that infection is the active agent in the spread of the disease. After exposure, it may show itself in a few hours, or during a period of ten days, possibly longer Two cases have just come to my notice, where the children of Mrs. L—, after exposure to infection, and being sent away from the house where it occurred were attacked; the disease showed itself in one child in twelve, and in the other in fourteen days.

During an unknown portion of this time the disease may spread to others, and it may be spread during the whole period of desquamation, and probably some short time longer. The disease seldom occurs twice in the same individual, so that those who have once had it are not likely to suffer again if exposed to infection. There is no disease so infectious. It may be conveyed by articles of clothing. On this account medical men and others should use every precaution to prevent such an occurrence. [1] Walking some distance in the open air after leaving the patient's house, instead of driving; changing a coat if it has touched the patient, washing the hands, especially in a little Condy's fluid mixed with water, and avoiding touching other patients till some time has elapsed, and making the visit last instead of first in the morning round, are the precautions that are used and found sufficient. It is believed that a temperature of 212°, also the fumes of nitrous acid, will destroy the virus.

Burning sulphur is also employed, washing and baking clothing, whitewashing rooms used by scarlet fever patients, and removing the paper.

During the time of illness, hanging up clothes wet with a solution of carbolic acid; placing Condy's fluid, in vessels about the room, the covering up and removing of the excreta as speedily as possible, the isolation of the sick person, and free ventilation are the proper measures to adopt. At the termination of desquamation a warm bath should be given every night for three or four nights. The fewer people that go into the sick room the* better. In the case of schools it may be needful to break them up, but it is a very unfair thing to send children home loaded with infection to convey the disease to others. It must also be remembered that infection can be conveyed in a box of clothing.

Last year I attended a little boy in scarlet fever, whose sister remained in the house, but was kept on a different landing, and every precaution was taken to prevent the spread of the disease. The little, boy got well and left home. His room was thoroughly renovated; he returned home after Christmas. Towards the end of March the little girl was attacked with scarlet fever. She had not been out of the house for thirteen days before her attack began. The problem to be solved is, how did she get it? Some unsuspected article of clothing might have retained the infection, or she might have taken it from some source outside the house. I must not take up time by going at length into this question; the mere remembrance of how plants are impregnated, and how minute particles are conveyed through the air, ought to help us to a solution of the difficulty. One other question. Will scarlet fever ever arise spontaneously? I would at once say no. "The disease maintains its type." The wheat found in the case of an Egyptian mummy has not lost its vital germ. Place it in a proper soil, and it brings forth "first the blade, then the ear, after that the full com in the ear" (Mark iv. 28). So the germs of these exanthems, they exist, and under circumstances favourable to their propagation, we may have an isolated case, or an epidemic.

Treatment. — Having considered at some length the varieties of scarlet fever and its complications, we have now to deal with the important subject of treatment. Those who have witnessed the severe forms of the disease will know the difficulty of the task. Our allopathic brethren have experienced this, and if we speak truly we must admit our difficulties, and lament over the disappointments we have met. The issues of life and death are not in our hands. God places means within our reach, and the ability to use them. We must avail ourselves of them, the result is with Him.

A history might be written of the various modes of treatment that have been adopted. Bleeding, purgatives, stimulants, antiseptics have all had their turn. It was, I believe, Sir Thomas Wateou who recommended chlorate of potash. His admirable lecturer took the place of Gregory's practice of medicine which furnishes a point of departure in the treatment of disease, and fairly represented the practice of medicine up to a little after 1840. Dr. Alison, Dr. Aitken and others succeeded, and led the way to what goes on in 1876, Are the results very different? In the *London Medical Record* of January, 1876, I find Dr. Brackenridge recommending the sulpho-carbonate of sodium. This, if successful, is to take the place of the chlorate of potash and other remedies, that each in turn have been recommended. The mention of these affords me the opportunity of pointing out the error that is committed in thus treating disease.

An allopath, from careful observation or from reasoning, is led to the use of some particular medicine; he meets, let us suppose, with a fair amount of success; cases similar to those he has previously seen fail, do well with his new remedy. He writes an account of his cases, other men try the medicine, perhaps with success; a few months pass over, the medicine begins to be less useful, ultimately it goes out of fashion; other men write about the complaint, and tell of the failure of the once loudly praised drug. Now, why is this? The disease has the same name at each period, but the symptoms are quite different. Surely a child may see the difference between scarlet fever of the malignant type, with a suppressed eruption and the same disease in a mild form, with a well marked eruption; need there be any wonder that the drug that is of use in scarlet fever of one variety Is perfectly useless in scarlet fever of the other? In "the good old days," when the doctor ordered number one ward to be bled, and number two purged, it is possible that in a few cases good was done and in others injury; so, if we attempt to treat a disease with one remedy, we must expect many failures. This has been tried with some homoeopathic remedies, it is certainly not homoeopathy. If in any particular epidemic we are fortunate enough to hit on the medicine suitable for the state of things then existing, the one medicine may be given with advantage in case after case, but no judicious homoeopath would act thus blindly unless he saw that there was a similarity in something more than the name.

In Asiatic cholera Dr. Quin gave camphor to a large number of cases, where the symptoms would naturally present very much the same features, and met with a success that has, I believe, never been approached by allopathy, and to this day we place great reliance on camphor; but it is used along with other medicines, as each seems indicated, and it might so happen, if we should unhappily be again visited with cholera, that some other medicine would be vastly superior to camphor. Even in the case of common catarrhal cough, the time of year, the existence of one or two fresh symptoms may make the medicine that is invaluable at one time comparatively useless at another.

Thus, without for a moment undervaluing the skill or unwearied industry of my brethren of the old school, I cannot but feel that they lose opportunities by failing to notice these points. I am far from saying they are always overlooked, but they are not looked to as they ought to be. In homoeopathy the great secret of success is to select a medicine that has, when given to persons in health, produced symptoms similar to those that the patient we are about to treat is suffering from.

I would here remark that as physicians, every mode of treatment that will secure success either in curing disease or lessening suffering is open to us, and the man that ties his hands so as to shut himself out from these means acts Unwisely towards himself and unfairly towards his patient; but it is from experience of various modes of treatment that those who profess to practise homoeopathically have come to the conclusion that as a general rule in using that method of treatment they are doing the best they can for their patient, while they still give the higher law the pre-eminence, being ready to employ palliatives or other measures if they see need. Thus, if I have a patient suffering severe pain or want of sleep that I fail to relieve with what I consider to be properly chosen homoeopathic remedied, my duty clearly is to procure the relief my patient needs by the best available means in my power. The homoeopath who the least frequently has occasion to resort to such measures is the safest practitioner, and he who is frequently resorting to them probably understands but little of homoeopathy, and in all likelihood when he has tried what he calls homoeopathy has in reality been treating a name, rather than the whole group of symptoms presented by the case, which, after all, is the only index to guide us to a light selection.

Treatment of *Scarlatina Simplex.* — To come now to the disease we have to deal with. A child is out of sorts, there is probably headache, loss of appetite, chilliness with fever at night, the skin being dry, hot, and burning, there is thirst and restlessness. There may be no rash. Now, under these circumstances, it will be well to give the patient *Aconite.* This should be given every two or three hours, with directions to give it about every half-hour at night till sleep is, procured, if the patient is wakeful and feverish.

But suppose there is in addition inclination to vomit or absolute vomiting, which is very likely to be the case, *Aconite* will still be sufficient, unless it should happen that vomiting and nausea are the leading symptoms; if after some doses of Aconite there is no change in these symptoms, then a dose of Ipecacuanka may be given with advantage every ten or fifteen minutes for three or four doses.

In giving homoeopathic medicines, I mean in small doses, you cannot be sure what channel nature will act through. Thus, *Ipecacuanha* may secure relief by inducing free vomiting, though you have not given it in sufficient quantity to secure its emetic action, nor had you any such object in view when administering it; you simply gave it because you know its action on persons in health is to produce nausea and vomiting; therefore when those are present it ought to relieve them. It may, and in most cases it does, secure the desired relief by the quiet subsidence of the discomfort. Should it appear that vomiting is likely to ensue, there is no objection to resort to the old method of aiding it by allowing the patient to drink freely of warm water, if old enough to understand your wishes, and if the state of the throat does not make swallowing too painful.

As soon as the necessity for the *Ipecacuanha* has gone by, the Aconite may be continued as before. Another use of *Ipecacuanha* that I may mention now, though I will have to allude to it again, is its value in cases of suppressed eruption.

In some cases the throat is one of the first symptoms complained of. When this is the case there certainly is no medicine more frequently needed than *Belladonna*. The pain in the head, the feverish symptoms, and with these the very probable existence of some delirium at night, or if the case is mild an approach to it, all lead to the selection of this medicine, and though *Aconite* is still indicated, it only becomes the primary medicine when the high fever, hot, dry, and burning skin, for the time take the prominent place. This is likely to happen in the early part of the night. The accession of fever, as evidenced by the rise of the thermometer in the evening, will be an inducement to give the *Aconite* at that time. Thus we may, perhaps, give *Belladonna* every two hours during the daytime and towards morning, but *Aconite* every half-hour at night, if the patient is wakeful and feverish.

In those cases where we may be hesitating between *Aconite* and *Belladonna*, it will be well to give three doses of one, then three doses of the other, with directions to give one or more extra doses of whichever of the two medicines gives most relief. Where it is necessary thus to trust a little to the discretion of those in attendance, always leave explicit directions; written down is safest. If it is found that the attendants fail in discretion, withdraw all such provisional power, and watch the case more closely; it is better that you should have a little more trouble, than that the patient should suffer from ill-carried out directions. By all means secure one, or, if necessary, two good nurses; their value cannot be measured by their cost in money. And let me say, to the credit of our nurses at this hospital, [2] that those in the house, as well as those who have been trained here but are now gone from it, are equal to the average of any nurses in London.

Once the rash has come out, the indication for *Belladonna* is much greater, as amongst the symptoms produced by this medicine we have "redness and heat of the whole body," "cutaneous eruption resembling measles," "dark red scarlet-coloured spots on the whole body," "scarlet eruption (the first days)." You will remember that in homoeopathy it is "like cures like," not the same cures the same; therefore, if we are told that Belladonna does not produce scarlet fever, no one said it did. But we do say it produces symptoms very like it; therefore we expect it to help to cure it.

Swelling of the glands is another indication for this medicine, and as this is a very common symptom, it will strengthen our confidence in the medicine we have chosen. As auxiliary means, a potato baked in its skin, then wrapped up in a handkerchief and squeezed, will be found an agreeable substitute for a poultice; the handkerchief prevents it coming in contact with the skin, but it moulds itself to the shape of the part to which it is applied: it will retain its heat for half an hour or an hour, and can be renewed as often as desired. It is a grateful application in all cases of sore throat. If care is taken to avoid a chill, sponging the hot burning skin with warm water is very grateful. If a dry sheet is placed underneath first one part, then another, so as only to leave a small portion uncovered at a time, the whole or greater part of the body may be washed over. If the headache is very bad, as it is sometimes in the early stage, the patient may prefer being let alone; avoid worry, and as far as possible let what is done be soothing, not irritating.

If the case continues mild, no threatening of dropsy or suppuration of glands, medicine may be given less frequently — a dose every three or four hours being sufficient.

When the skin is coming off great care must be taken not to expose the patient to any chill. It seems a needless precaution to a foolish mother, or a wilful child, to keep the patient a good deal in bed during the period of desquamation, if it is eating well, and is apparently quite strong. Rest assured it is a wise precaution, the child will not suffer, and "will be kept out of harm's way." As strength returns it may be wrapped up in a blanket, or warmly clad be allowed to sit np for a short time, daily. It has been proposed to oil the body of the patient before the stage of desquamation, and this has been done; it is said that it prevents the peeling off of the skin, and so to some extent stops the spread of infectious matter, besides being grateful to the patient, and preventing albuminuria. I have not adopted this practice, and merely name it as a measure that has been resorted to. So far as the patient is concerned, regarding the desquamation as a natural process, I doubt very much the advisability of checking it, though the argument that so doing helps to prevent the spread of infection is not without value.

Towards the close of the desquamative stage it is desirable to give a few doses of *Sulphur;* a dose three times a day for three or four days may be given. This is a medicine of searching action, acting well in scrofulous affections, in cases where there is emaciation, in skin diseases, and exercising an influence on almost every organ of the body. Under these circumstances you will find an advantage in giving a few doses of it at the close of every exanthematous disease. If there is anything lurking in the system it will help to dislodge it; in some cases it will cause an aggravation of symptoms, but though there may be temporary aggravation it is usually of a curative character, for as it subsides the disease may yield, or some other homoeopathic medicine that may be indicated will act much more effectually after the *Sulphur.* When we read of the following among the indications for the use of this medicine, we will see how useful it may prove in many affections of the skin, whether acute or chronic: "Eruption on the skin," "Fiery scarlet eruption over the whole body," "Scurfy eruptions with red areola and itching," "Dry scaly eruption."

For the miliary eruption that may show itself with the mild form. *Belladonna* will still be of use. Another medicine also deserves notice, the *Ailantus Glandulosa.* This medicine is strongly indicated where the miliary eruption is present, but otherwise seems more suited for the anginose variety, or where the eruption is more dusky than florid, than for the mild form, when Belladonna is the best medicine.

In what may appear a simple case, if more serious symptoms threaten, the medicine needed for this will find its place more suitably under one of the other varieties — the treatment of which will now come under our notice.

Treatment of Scarlatina anginosa. — This variety, as we have seen, is characterized by the severity of the throat symptoms. At first the treatment may be the same as what has been recommended, and if the case appears to be doing well there is no need to alter it, the state of the pulse, the temperature, as ascertained by the thermometer morning and evening, the debility, state of head, delirium, and the appearance of the throat, guiding us as to how far we may safely trust to *Belladonna*. Amongst its symptoms we have dryness in the fauces and burning of the tongue, inflammation of the throat and fauces, phlegmonous, with violent fever, inflammation of the velum pendulum, of the uvula, sore throat, stitches in the fauces, internal swelling, sensation of a lump, inflammation and suppuration of tonsils, inflamed parts covered with a white tenacious mucus, impeded deglutition. The high fever, eruption, and delirium are, as we have seen, met by this medicine, and no doubt it alone will be sufficient for many of the cases we meet. But not for all, and it is here, in the wise selection of another medicine, that we may benefit our patient. If divided between *Belladonna* and some other medicine, we may give three or four doses of the first, then three or four of the latter, or the *Belladonna* by night, to quiet the delirium, and the other by day.

You will be told in books to alternate your medicines, it is done by gentlemen of the highest standing amongst us; but as it is my business to direct you as to the best mode of practice, I would advise you not to adopt this method, it is contrary to the teaching of Hahnemann, contrary to what I may call the common-sense teaching of homoeopathy, which is to give only one medicine at a time; therefore, if you are tempted to alternate in any way, let it be as I have suggested: first three or four of the one, then three or four of the other.

I have spoken of *Ailantus glandulosa* as likely to be of use where there is miliary eruption, let us look at it now a little more closely as a suitable medicine for other symptoms.

Giddiness, nausea, vomiting, and headache, with flushings, heat, and heaviness of head, are met by this medicine; eyes suffused and congested. These are some of the early symptoms. Throat inflamed with ulcerated spots, tenderness of the parotid and thyroid glands, thickened and swollen feeling of muscles of neck. According to Dr. Chalmers's observation, "the throat is livid and swollen, the tonsils studded with numerous deep, angry-looking ulcerations, from which a scanty foetid discharge exudes; the neck is very tender and swollen." In addition to this Dr. Chalmers speaks of its use with copious thin ichorous discharge without fetor, and discharge of blood and pus in scarlet fever. These observations are gathered from "its use in disease."

Though wanting the value of an original proving, the repetition of success in such cases imparts a clinical value that is very important, and justifies the use of the medicine. Under such circumstances it may be expected that these symptoms will be found to be corroborated by those who prove the medicines.

The smallness of the pulse, drowsiness, insensibility, muttering delirium, dark miliary rash, almost of a livid colour, then again, as given by Dr. Chalmers, the scanty, patchy, evanescent, and often long-delayed rash, the presence of petechia?, and, as he says, the eruption being slow to make its appearance, and never taking on the genuine appearance of scarlet fever, but remaining of a livid colour, are just the sort of symptoms that we want to relieve, and as the medicine has been well recommended, we may hope that further experience will prove it to be one on which to rely.

There is a medicine that I am in the habit of using as a gargle in diphtheria, in the strength of five drops to the ounce — the *Phytolacca Decandra.* I have much confidence in it, and think that it might be employed in a case of scarlet fever in any way complicated with diphtheria, and supposing the child could gargla I should not depend on it alone, for it in no way covers the symptoms of scarlatina, but as an auxiliary it might be used two or three times a day, using it about an hour after the regular medicine, so that it should not interfere with its action. If the child cannot gargle, the throat might be brushed over with the lotion. The symptoms it would meet are — sore throat and swelling of the soft palate in the morning, with a thick white and yellow mucus about the fauces, the throat feels very dry and sore. *Hepar Sulph.,* or *Mercurius Iodatus* may be given internally.

Another medicine, covering a larger range of symptoms and more generally useful, and which has its advocates with allopaths as well as homoeopaths is *Ammonium Carbonicum.* In the *British Journal of Homoeopathy,* vol. xxiv., page 179, an account is given of an outbreak of scarlet fever where 192 out of 440 inmates of an asylum were attacked. Of these, 65 cases were simple, 78 anginose, and 49 malignant. All were treated by being given five grains of sesquicarbonate of ammonia in one ounce of water every four hours. Diet: milk, beef tea, and wine liberally when depression was present. Ten died, 9 of them idiots. Fourteen of the malignant cases occurred among the officers and servants, only one of whom, who had a diseased heart, died. Dropsy with albuminuria occurred in 12 of the cases.

This result is not unsatisfactory, as it is fair to suppose that when so many as 49 cases were malignant the anginose ones would be severe.

Of course carbonate of ammonia was not given in these cases on account of its homoeopathic action, but as a stimulant that would sustain the system during the continuance of the fever, and so bring the patient to a period of safety. Still, as the medicine would be homoeopathic to the disease, the specific effect, though lessened by the largeness of the dose, would not be lost altogether. In treating the disease homoeopathically, the same medicine would be given but in much smaller doses, as the stimulant effect would not be sought; indeed it is a question whether there is not a positive loss in giving a medicine in a quantity that goes beyond what will produce more than the homeopathic action. When the dose is too large it is positively hurtful. The tendency of some professedly homoeopathic practitioners to give medicines in doses that almost rival, or in some cases surpass, the old school in magnitude, is greatly to be deplored, as it strikes at the root of the teaching on which Homoeopathy has risen to the high position it occupies: it retraces the steps to the abuses of old physic, and shakes confidence by appealing to a natural prejudice in favour of substantial doses, while it is opposed to the experience of the best teachers of homoeopathy. What is really wanted is the curative, without any approach to health disturbance. Thus for example, the hundredth, the ten thousandth, or smaller quantity of a grain of Ipecacuanha may cure vomiting, but if I give fifteen grains I produce vomiting. I produce a violent effect, perhaps aggravate the mischief, without doing good, and lose the real curative action which the small dose would have possessed.

Giving the *Ammonia* in homoeopathic doses, the symptoms to which it corresponds are — "the whole upper part of the body is red, as if covered with scarlatina, the sleep is restless and unrefreshing, feverish chilliness or heat, headache, throbbing, and muddled state of the head; sore throat, raw feeling in the throaty lassitude, and weariness."

This medicine I would recommend early in the disease in cases that are beginning to assume a more grave form. Three or four doses of it after three or four of *Belladonna* can be given. It also may be of use when the patient is low, and when there is depression. It will be well to give it before or after a few doses of some other properly indicated medicine, or alone if sufficiently well indicated. It is a medicine that may be thought of in cases where there is a drowsy, muddled state, as if the system were oppressed with the poison of the disease, or under such conditions as a useful adjunct to *Opium,* of which I hope to speak presently.

One of the most important medicines supplied by the animal kingdom is *Apis,* the common honeybee. The headache and confused state of patient, the retrocession of eruption, the extreme heat, injected eyes, congestive state of head, and violent delirium would alone make this a useful agent in many severe cases, but it is further indicated by the raw burning feeling in throat, scalding and soreness of tongue, painful blisters, sensation of contraction, and difficult deglutition, copious accumulation of soapy saliva, burning, stinging sensation in mouth, fiery red appearance of the buccal cavity, with painful tenderness. Various kinds of angina with redness, swelling, and stinging pains.

Apis produces erysipelatous-looking swellings, with oedema, so that in this medicine we have one that we may look to with confidence in some of those cases ushered in with a sharp outburst of sore throat. It is also a medicine that will help us in the dropsy that may come on later. When speaking of this, it will again be necessary to allude to *Apis* as one of our best medicines.

Another remedy furnished by the animal kingdom is *Lachesis*. This will help us in the anginose as well as the malignant variety. Besides the scarlet-like eruptions, swelling of the cervical glands, black appearance of lips and tongue, feverish symptoms generally, moaning, excitement, small quick pulse, delirium, and loss of consciousness, the throat symptoms are well marked, and show we have a valuable remedy for the advanced stage as well as earlier. We have burning in the throat, sensation of rawness, inflammation of the throat, and of the tonsils, with redness and disposition to suppurate, phlegmonous inflammation with violent fever, ulcerated places in the throat, inability to swallow, gangrene of the tongue, difficulty of speech, great sensitiveness of the neck and sensitiveness to contact, red suppurating blotches under jaw, turbid dark urine, difficulty in urinating.

Now, having treated a case with *Belladonna* and *Ammonium Carbonicum*, or with *Apis,* if better indicated, as the disease advances we look to *Lachesis* as supplying a remedy that covers a large group of symptoms. A case I attended lately that was otherwise doing well became complicated by tenderness and swelling of the glands of the neck; *Lachesis* gave relief. One other peculiarity about this case was a large patch of dusky redness on the forehead that gave me some anxiety, but this disappeared without trouble. *Lachesis* was suitable for this symptom.

Hitherto the medicines spoken of are those that most nearly represent the disease in the mild and anginose forms. Of course there are others that may be of use, and that are often needed for special symptoms. Before speaking of the most important of these, I propose speaking of those called for in the malignant form of the disease, and in cases without eruption, taking up afterwards the medicines needed for special purposes either connected with the earlier or later stages. In thus dealing with the subject the student of homoeopathy should remember how impossible it is for the teacher to do more than give a general outline of his subject. In each case, if difficulty arises in the selection of a remedy, or if any peculiar symptom should not be met by any of the medicines named, then a reference to the repertory is indispensable, and rest assured that, however much some short road to homoeopathic practice may be recommended, the safest and best practitioner is he who by the aid of the indications given in his repertory chooses his medicine.

Treatment of Scarlatina Maligna. — As in the former varieties, so in the malignant, *Belladonna* will be found of much use. Besides the sore throat and other indications spoken of, the presence of stupor, starting in affright, wild look, earache sleeping with the eyes half open, a strong and quick or full and slow pulse, suppression or retention of urine, involuntary urination or stools, or convulsions will be indications to guide us.

Rhus Toxicodendron is a useful medicine in most varieties of fever, and does not fail us here. Some newer remedies have been more in fashion lately, but this is one of our old tried remedies that will, I trust, not be lost sight of. Rheumatic pains are prominent guides, if such should be present The usual chills and heats of fever, miliary rash, pulse quick, or slow and irregular, anxiety, sore feeling of nostrils, red face with heat, the lips dry and parched, covered with a reddish brown crust, parched, red or brown tongue, discharge of pus from ears. The throat symptoms are not so marked, but there are glandular swellings* Petechial spots with great debility. Retention of urine, incontinence of urine. Small burning vesicles with redness of skin. Ulcers as if gangrenous from small vesicles, with violent fever.

In a prolonged case of malignant scarlet fever, or in a continued fever following the eruptive fever, or in erysipelas, which shows itself occasionally, *Rhus* will be found a most useful medicine sometimes alone, sometimes given for three or four doses after each three or four of *Lachesis, Ammonium Carbonicum,* or *Arsenicum.*

It may be well to refer back to what has been said earlier about *Lachesis,* and also about *Ammonium Carbonicum,* and *Ailantus,* as they are medicines that may be of me in various stages of the disease, whether of the malignant or anginose variety.

When we are compelled to leave our patient for twenty-four hours, the plan proposed earlier has many advantages; if the medicine first given is doing good, directions should be left not to change it; our knowledge of the natural course of the disease will enable us to anticipate a little what may be needed. In this way we can sometimes gain the advantage that would be derived from an additional visit — a matter of some importance when a medical man has to carry on a large and scattered practice without help, and in many cases it is an object to the patient to be spared the additional fee. Still, if possible, the truest practice is to give but one medicine, watch its effects, and decide from actual observation when it should be changed, but remember the fewer changes the better. Select your medicine with care, and do not change it for light reasons.

Arsenicum is indicated by an ulcerated condition of nostrils, with acrid discharge, or dryness; epistaxis, sunken, pale, death-like, yellow, livid countenance, or bloated puffed red face, ulcerated eruption about lips. Tongue cracked, brown, blackish, parched, trembling when put out. Foetid smell from mouth. Aphthae. Burning in throat, gangrenous sore throat, swelling of sub-maxillary gland. Loss of consciousness; staring eyes; wandering. Paralysis of bladder. Restless sleep, startings. Petechiae. Chilliness or violent burning heat, putrid ulcers. These symptoms are sufficient to show what a valuable medicine *Arsenic* is, though the eruption may not be so characteristic as that of some other medicines. When the skin is peeling off, as also when there is dropsical swelling, this medicine is of use. I do not, however, think *Arsenic* equal to *Apis* in the dropsy of scarlet fever, though in cases where there is great debility, or where there has been much putridity, it may be preferable.

Suppression of Rash. — In some cases we find that from the very first the patient is struck down as if from a mortal blow, the dose of the poison has overpowered everything. There is no rash, or it is of a dusky colour, or it comes out badly and then fades away. The patient is in a drowsy state, is roused with some difficulty, and falls off to sleep again. On examining the throat we may find comparatively little to be seen, or it may be of a dusky hue.

Case 1. — I recollect a fine little boy whose parents had not taken alarm, probably from the absence of the urgent symptoms that they associated with the disease in their own mind. They had not sent in the first instance; and when I saw the child, the drowsiness and stupor were so marked that I thought it very possible the case would be a short one, and that the child would succumb to the overpowering poison of the disease. Under these circumstances I should, had I been practising as an allopath, have trusted to rouse the patient by applying sinapisms and giving carbonate of ammonia; and I do not think that such treatment would have been bad; but practising homoeopathy, I had a fixed rule to guide me, an unspeakable comfort under such circumstances. I knew that I had in *Opium* the remedy that I wanted, and I administered it in rapidly repeated doses— observe, not in large doses. The object was gained for the moment, the child was roused and passed through the first stages moderately well, but never seemed entirely to shake off the complaint; secondary fever set in, lasting about ten days, when he sank.

Opium. — Having tested the value of this medicine at other times, I can speak with confidence of its powers; and though it is the one that would first come into my mind, yet I would remind you that there are other medicines, such as *Hyoscymaus* and *Belladonna* that, guided by your repertory, you maybe led to give a preference to. In these cases the brain seems as if overpowered by a narcotic poison; therefore amongst drugs producing this action the homoeopathic remedy should be looked for. In such a case as the one I have spoken of, a few doses of *Carbonate of Ammonia* may follow the *Opium* with advantage.

Case 2. — Master G. R, set. 8, was observed to be a little out of sorts for a day or two. On April the 15th, 1876 he went out for a walk early in the day. As the evening came on, the slight appearance of illness before observed became more marked, and he was attacked with vomiting, as I was informed.

April 16. — Having had a restless night, and appearing to be very ill, I was sent for. I forwarded directions that he should have a dose of Aconite every hour till I saw him.

On my arrival I found the child lying in a heavy sleep, snoring, the noise being as it appeared partly due to the state of the throat, and partly to the oppressed character of the sleep. The face was congested and livid. The night had been a disturbed one. On rousing the boy there was incoherent answering; thus in reply to a question as to whether he had pain in his throat, he began to spell k-n-i-f. At times a more collected answer was got from him; but at others he was confused. Asked where there was pain, he pointed to one of his wrists. Later in day he did not recognise his father.

On looking at the throat the tonsils were observed to be swollen, but not red; a quantity of white mucus was pressed up over them, but the appearance did not at all give the idea of much soreness, nor could we understand from the child that there was. The tongue was coated with a thick whitish slightly brown fur. The pulse was about 136. On examining the back and abdomen I could detect the appearance of a somewhat dusky rash; this disappeared under pressure, but immediately returned again.

The case was distinctly one of scarlet fever, with a partly suppressed eruption, the lividity and congested look of face, with the drowsiness, stupidity, and character of the sleep, all pointed to a greatly oppressed brain — oppressed as if from some powerful narcotic poison. There was no excessive heat of skin, and, as already mentioned, the throat symptoms were by no means marked.

The danger was considerable; if relief were not obtained speedily the child would probably die, poisoned, as it were, by scarlatina poison. Under these circumstances I had to think of what remedy most nearly met these symptoms. I ordered *Opium* 12x every hour for four doses, then *Belladonna,* three doses — then four of *Opium,* and so on; giving directions to keep more entirely to the *Opium* if the drowsiness continued; but if it abated to give less of it and more of the *Belladonna.*

April 17. — I was greatly gratified on making my visit to find the livid countenance had given place to what was now a red, still somewhat congested one, with redness of eyes, some blood on teeth induced by picking at lips. The eruption was out much better over the body. He answered questions, wished for something sweet, the sleep was less marked by the snoring, now due to sore throat alone, not complicated, as it appeared to be yesterday, by the narcotic snoring. The tongue was still coated, but less of the brownish tinge. The throat showed some patches of ulceration, not much complained of. The child had passed water when taken up for that purpose. Pulse 128; temperature 101 ½.

The immediate danger was now over. The further history is merely of interest as showing the progress of an ordinary case.

Four doses of *Belladonna* were now ordered to each two of *Opium.* The intervals to be lengthened to 1 ½ hour.

Diet — milk and barley water, and beef tea.

April 18. — Had a good night, slept four hours at one time.

Eruption out, tongue cleaning a little, throat better. Pulse 120; temperature 1021. Asked to be taken up to make water. Continue medicine.

April 19. — Rash still out on body and legs. Slept well. Pulse 100. Temperature in night was 100.1. Tongue shows red raised papillae. *Bell.* 3 every two hours.

April 20. — Had a good night, except for a couple of hours in middle of night, when he was disturbed with a dry cough. Voice still thick; otherwise doing well. Bowels have been comfortably relieved, after being without action since attack began. Pulse 100. Rash nearly gone.

Bell., three hours. *Hep.* 12x, two doses evening.

April 21. — Very little trace of eruption on legs. Going on well, but was again disturbed by cough in night. Uvula is long.

Hepar, four doses in day. If cough at night, *Kali Carb.* 12x every half-hour or hour.

April 22. — Cough greatly relieved. Continue *Kali* as need.

April 23. — Desquamation commencing.

May 8. — Doing well; skin still coming off.

May 15. — Going on well; bowels rather confined, skin still coming off.

In some cases where there is no rash, or but little sore throat, if there is an absence of any symptom of danger, all that will be necessary is to watch the case, and take the ordinary precautions to prevent the spread of the disease.

In these cases the possible occurrence of dropsy at a later period must not be forgotten. The disease may escape notice, owing to its mildness, and yet Ufe be imperilled or lost by the after mischief.

Suppression or receding of the rash must always be looked on with anxiety if other symptoms show gravity. Such medicines as *Ipecacuanha, Opium,* and *Bryonia* may be used to help its development. A warm bath may also be of use. In all cases of fever remember your thermometer. Since we have got a registering instrument the difficulties of former days have vanished.

Isolated symptoms will give much trouble in severe cases, and as with Opium and the allied remedies for stupor or coma, so we must be prepared to use occasionally medicines that meet these special circumstances.

Convulsions may appear. The remedies we may use are *Belladonna, Nux Vomica, Opium, Ignatia, Cuprum* and *Veratrum Viride. Belladonna* will be of use where there is excitement, rolling the head on the pillow. *Opium* sleeping with eyes half open if awake muddled or unconscious, with picking of the bedclothes, picking at the lips. *Nux Vomica* where there are violent spasms. *Ignatia* where there is jerking of the muscles, sleeping with one eye open. *Cuprum,* convulsive movements of limbs, fits of an epileptic character. *Cuprum* may be of use where there is great muscular weakness, as well as under other conditions. *Veratrum Viride* in cerebral congestions and cerebro-spinal meningitis, facial twitchings and tetanic spasms, fever.

Enlargement of the glands, which may prove a serious source of trouble, especially if associated with cellular inflammation, requires careful looking to. *Belladonna, Mercurius, Lachesis,* may help us; where suppuration threatens *Hepar* should be administered. This medicine has long been of use in homoeopathic practice as a medicine to promote suppuration. As a rule I do not think its use is resorted to sufficiently early, the fact being overlooked that it will often arrest suppuration as well as promote it when that stage is reached.

Where there is chronic discharge from these abscesses *Silicea* may be used with advantage.

Dr. Sidney Binger, to whom all credit is due for availing himself of every source of information, is introducing remedies of this class to the medical profession. The fact of his gaining his information from homeopathic sources ought to gratify us, even though he may not yet see his way to acknowledge himself a convert to homoeopathy.

Silicea, Hepar and *Sulphur* may be given for a few doses where there is purulent discharge and soreness about the mouth or nostrils.

Treatment of Complications and Sequelae. — Diarrhoea must be met by such medicines as *Arsenicum* or *Veratrum,* or, if there are bloody stools with much tenesmus, by *Mercurius Corrosivus.*

Erysipelas will show itself at times. *Apis* and *Rhus* are the medicines most likely to be needed; the first when the affected part has a swollen, shiny appearance, as if from the sting of a bee, the other when there are vesicles; *Belladonna* when there are head symptoms, and *Lachesis* and *Arsenicum* when it assumes a gangrenous appearance

Pneumonia, pleurisy, jaundice, and similar complaints arising in the course of scarlet fever, are accidental circumstances not necessarily connected with the disease, so that it would be out of place to enter on their treatment here.

Continued fever following scarlet fever is seen from time to time, and will require such medicines as *Bryonia, Rhus, Arsenicum. Baptisia* is recommended in fevers of a typhoid character. I am, however, much inclined to think that the two I have first named will be more frequently needed.

Otorrhoea, which is such a common accompaniment, will be best met by *Pulsatilla, Hepar, Mercurius,* and *Silicea.*

Delirium is met by most of the medicines already named; but where it is of a very violent character, and has not yielded to *Belladonna* or *Hyoscyamus,* I know no medicine more likely to subdue it than *Stramonium.* This should be given in small doses every twenty or thirty minutes where needed.

The most common and most dreaded of the sequelae, dropsy, I have left to the last. Having already spoken of the symptoms that usher this in, it remains but to speak of the treatment.

As a precautionary measure, I have alluded to the advisability of keeping the patient in bed during the greater part of the period of desquamation. If allowed to be up, he should be confined to his room, for his own protection and that of others, as at this stage he may help to circulate particles that will produce fruit if they fail on a suitable soil. He should be well clothed, and if the weather is cold, protected by woollens or flannels. As the desquamation draws to a close, he may have a bath each night for three or four nights, of a temperature of 96 to 98, which must not be allowed to cool too much. If weak, it will be safer not to give the warm bath, but to be content with washing the skin, doing it bit by bit, to avoid unnecessary exposure to cold.

If albuminous and scanty urine and dropsical swelling appear, such symptoms should be met at once. The urine should be tested from time to time. As a rule, the medicines on which I rely are *Apis* and *Squilla*. The *Apis* for simple dropsy with albuminuria.

In the *British Journal of Homoeopathy*, vol. ii., p. 409, a case of a little girl aged nine, under the care of Dr. E. A. Munger, America, is given; she had gone through an attack of scarlet fever from the 14th to 16th December, 1851. Doctor called to see her on 31st. Face had been swollen three or four days; belly was full and hard; pulse was 160. Breathing laboured and panting; could not lie down. Urine scanty and high coloured; skin hot, dry. Thirst. Free from pain. Was treated with *Acon.*, and *Arsen,*, with some diminution of thirst and external heat. In the evening she had *Bellad.* and *Arsen.* On January 1st the limbs were cold. No increase of urine. Respiration more laboured, pulse quick and intermitting. She was given *Acon.* and *Helleb.* In the evening pulse was fuller and more regular, otherwise no change. Was given *Dig.* and *Helleb.*

Jan. 2nd. — Much worse. Face and limbs livid. Breathing gasping and excessively rapid; pulse indistinct; cold sweat; face and extremities cold.

Doctor did not expect her to live over day. He left *Apis* 3 — a drop to be taken every two hours. Was surprised to find patient alive and better in afternoon. The livid colour of the face was gone> a slight warm sweat had appeared, the extremities were warm, the respiration better than it was at first visit, although still laboured and accelerated; the pulse fuller and regular. Since noon she had commenced to pass water in large quantity, and it was still increasing. She continued with the *Apis* every four hours. In the morning of 3rd January patient was in every respect better, able to lie down the first time for four days. During the night she passed three pints of urine of a dark colour without sediment. The medicine was continued, and by the 10th she was well

Dr. John Wilde, now of Weston-super-Mare, read a paper in the British Homoeopathic Society, giving an account of an epidemic that he had to deal with in 1863. One of the cases he asked me to see in consultation with him, where there was dropsy, did well under *Apis,* Dr. Wilde has not seen the same results from this medicine that others have; he suggests the possibility of this being owing to the preparation he used. In the instance referred to I think it very probable that the *Apis* was supplied from my case.

Squilla is likely to help where there is smoky urine. In a case I saw in consultation with Dr. Eugene Cronin, of Clapham, this was a marked feature; there was also urinary irritation, a desire to pass urine, while that fluid was scanty. *Squilla* effected a very satisfactory cure. *Arsenicum* and *Hellebore* are very useful medicines, and should not be lost sight of. *Cantharis* and *Tere-*

binth may be called for in some few cases, but I think *Squilla* covers the symptoms we are likely to meet better. A paper appeared in the first volume of the *Annals* of the Hospital and Society, where Dr. Yeldham speaks favourably of *Terebinth* in suppression of urine, and no doubt h is a valuable medicine where it is fairly indicated. There are two medicines that I may say a word about — *Digitalis and Apocynum Cannabinum.* From my own experience I cannot recommend them as pure homoeopathic remedies. As powerful diuretics there is no doubt about their value, but when homoeopaths tell us they prefer *Digitalis* in the infusion, as it acts better, it is evident they are thinking of its diuretic action. I do not say we are to exclude ourselves from their use in this way. In a case of an old man who had been a large drinker, and who died of dropsy, and who was tapped once, but would not submit to this again, I gave these medicines expressly to act as diuretics when the time appeared to arrive for their use. Nor do I see why I should not use them as well as the trocar; but if you can cure your patients homoeopathically, as I believe you may hope to do in the majority of curable cases, it is, I think, unwise to lay aside the simple measures that are placed within our reach, and resort to allopathic practice. In organic disease you may only be able to palliate, but in curative diseases trust to the right remedies.

In former days, when patients were kept smothered up in bedclothing, fresh pure air being carefully excluded, the chances of recovery were certainly lessened.

In the present day a wiser and more humane method gets rid of these barbarisms. We are not afraid to let the patient have some fresh air, though this must be done without allowing the cold air to blow straight on him. An open window on a staircase or in an adjoining room may be found to answer. A pleasant, even temperature must be kept up.

So also when we see how grateful a sip of cold water is, it seems but rational to allow its use in small quantity, though we may forbid a long draught, as a large quantity of cold water suddenly thrown on the stomach might cause pain or even serious mischief, as it might do by suppressing a rash.

The patient is not in a state to eat or swallow much, but with beef tea, beef juice, milk; &c., the strength may be kept up.

If stimulants are needed, they should be given without hesitation. Delirium will at times be relieved by the judicious administration of stimulants; indeed I believe it is often an indication for their use. They should be given in measured quantity, and the effect watched. I believe they are often the means of saving life.

All offensive matter should be removed from the room as speedily as possible, as well as superfluous furniture and bed-hangings. Condy's fluid in the room and carbolic acid solution in the passages may be used freely; and lastly, when the patient is well, a thorough process of deodorizing, whitewashing, papering, &c., should be gone through.

Preventive Treatment. — The name of the illustrious Hahnemann is so associated with the use of *Belladonna* as a prophylactic, and the subject itself is of such vital importance, that the evidence for and against this must be carefully weighed.

Allopathic writers allude to the statement of Hahnemann, but usually in a way to discredit it. However, it was from one of these writers that I learned how to administer it. and when practising as an allopath I gave the extract mixed with cinnamon water, and acquired much confidence in it. As a homoeopath I always employ it, though usually in the shape of the tincture. I have seen it apparently successful in preventing the spread of the disease. I may say on most occasions, but I have seen the disease appear where the *Belladonna* had been given. It is very difficult in these cases to know whether the disease was not latent before the prophylactic was used. In one case, about which there could be no mistake, the *Belladonna* failed.

The evidence of homoeopaths is in favour of *Belladonna:* it is thought either to prevent the disease or to make it milder.

Dr. George Wyld, in the fifteenth volume of the *British Journal of Homoeopathy,* has collected the evidence for and against. He says the evidence is greatly in favour of Hahnemann; out of 2,027 which took *Belladonna* (Bayles' Catalogue), only 78 took scarlet fever of those exposed to infection.

Dr. Wyld thinks that Belladonna fails in malignant scarlet fever both as a curative agent and prophylactic. If he has trusted to *Belladonna* alone as a curative agent it would probably disappoint him. I have endeavoured to show you that in certain stages other medicines are more to be relied on in the malignant form of this disease.

Dr. Wyld advises the looking for other prophylactics. It must be remembered that even when no Belladonna is given the disease may not spread, so that while giving it its due we must be careful not to overrate its value. I should always recommend its use.

Dr. Joseph Adams, Physician to the Small-pox Hospital in the early part of this century, seriously discussed the propriety of exposing persons to the infection of scarlet fever during the existence of a mild form of the disease (Adams on Morbid Poisons, 2nd edit., 394). Such a proposal is more curious than valuable, as it is impossible to predict what form the disease will take. A severe form of the disease may show itself, even when contracted from a very mild case.

Measles

Measles, before the time of Sauvages in 1768, was known as Morbilli. He called the disease Rubeola, which has caused confusion without any good arising. The Arabian physician Rhazes first described it about the year 910 (his works, as well as those of his follower Avicenna, form one of the volumes of the old Sydenham Society). How much earlier the disease existed we can only conjecture. The confusion that existed about these diseases before Sydenham clearly separated measles from smallpox was very great. He laid the foundation for clearer views, but the complete separation of measles from scarlatina was not effected till a later period.

Measles is conveyed by infection from one person to another. Whether I am right or wrong I cannot say, but I have always believed it was less likely to be carried than scarlet fever. This latter disease may be conveyed by a person who has been in contact with some one suffering from it to another who has not otherwise been within reach of infection. It is doubtful if measles is ever thus spread. Still, as it is no great trouble for a medical man to wash his hands, or be a little in the open air before going fresh from a patient with measles to some other case, the precaution can do no harm and may save the possibility of risk.

It is said to require a period of incubation or latency of from six to thirteen or fourteen or even sixteen days. This is longer than my own experience would lead me to fix, ten days being the time in which I think it is most likely to show itself. If asked the question, how soon may a child be considered safe after exposure to infection, by naming a day or two over the longest known period of incubation, we will guard ourselves against error, but it is more satisfactory to be able to predict the probable appearance of the disease on a given day.

Measles is almost always ushered in by catarrhal symptoms, men we see a child suffering from cold in the head, sneezing, running from the nose; suffused or sore eyes, with swelling, redness, or irritation of the lids; hoarse, irritative cough, there is a strong suspicion that measles is likely to follow; while if in addition there are rigors, headache and flushing, nausea and vomiting, furred tongue, quick pulse, fever, and general appearance of illness, the attack may be a smart one.

Generally the fourth day after the appearance of this primary catarrhal fever, but sometimes on the third or fifth, or even later, a number of small red spots begin to make their appearance; these are round like fleabites, commonly of a deep red, but varying in colour, in severe cases assuming a more dusky or livid hue: these spots are slightly prominent, in some cases they become distinctly papular, sometimes vesicular; they disappear on pressure, but return on its withdrawal; they commence on the face, neck, and arms, spreading to the trunk, then down to the feet. A miliary eruption is at times seen. Measles has been mistaken for small-pox. The cough and coryza point to measles. The eruption of small-pox comes out forty-eight hours after the rigor; measles seventy-two hours, or later. Gregory says the papulae of small-pox are firmer, and are granular. At an early stage, no small amount of the *tactus eruditus* will be needed to note the difference.

We find papular eruptions in measles, strophulus, lichen, and prurigo.

The late Dr. Rutherford Bussell related some experiments of Simon's in the 10th volume of the *British Journal of Homoeopathy* which are of interest: — "Where tartar emetic was rubbed into an animal that was then killed, and the skin examined after the effect of the tartar emetic was produced, the cuticle was healthy. In the most superficial layer of the true skin, a small eminence about half the size of a barley-corn was seen. This little mound was caused by the injection of the part with blood. The fluid resembled the blood serum in consistence, and contained minute molecules and fat globules. In the immediate neighbourhood of the inflamed spot there was an unusual quantity of blood, and beyond this the skin was perfectly healthy." This description Dr. Bussell goes on to say, corresponds with that given by Heuld and Rokitansky, and differs from that of Rosenbaum, Hebra, and others, who conceive the sebiparous glands to be implicated in papular affections. The effused fluid may either be absorbed, and so on the subsidence of the injection of the vessels the pimple disappear, or it may be of so great an amount as to convert the papula into a vesicle, and this vesicle may in turn be converted into a pustule.

Simon cut out a pimple produced by measles and examined it under the microscope; he found it to consist of a spot injected with blood, but perceived no fluid such as was observed in those pimples produced artificially: however, the absence of fluid might be accounted for by the pressure employed during the operation, and it is most probable that the elevation of the spot depends upon the presence of an effusion. The cuticle adhered to the skin in a healthy manner, and there was no increase nor any unnatural appearance in the papillae of the skin. All the spots he examined were perforated by a hair, the ducts of which were healthy. He adduces grounds for disbelieving that the sebiparous glands are connected with this eruption.

The spots come out in crops. If they have appeared on the face and neck on the fourth day, the following, or fifth day of the disease, and second of eruption, a crop comes out on the trunk. On the sixth of the disease, and third of eruption, a crop comes out on legs and feet. The next day, the seventh, it begins to fade from the face, the next day from the trunk, and the following from the legs and feet. Thus after six or seven days the eruption is fast disappearing. A few rare cases have been observed where the rash has reappeared after having apparently completed its regular course.

The fever is at its height when the eruption is well out; after this, if the case goes on favourably, the temperature may fall rapidly. In some cases a temperature of 109 may be reached. In regard to the temperature, as might be expected, if this should be high during the catarrhal stage, it indicates a severe attack; but a high temperature coming on with the outbreak of eruption and subsiding with its decline is what may be expected.

The eruption of measles differs from that of scarlet fever in the larger size of the spots, their greater distinctness, greater tendency to come in patches, and the crescentic shape of some of these patches. The eruption may be followed by a scaly, mealy desquamation, but this is neither so marked nor so regular in its appearance as in scarlet fever; indeed, it may almost pass unobserved. At times there is itching. The eruption is the chief characteristic of measles, for the catarrh may be absent. The disease is then spoken of as *Rubeola* or *Morbilli sine Catarrho.* Though this is noticed by authors, it is not common. At one time it was thought that when the catarrhal (Symptoms are absent, the attack would not secure immunity if the subject of it should again be exposed to infection. About the year 1845 I attended a young lady in an attack of this character. The late Sir Henry Marsh speaking to me on the subject thought that these cases always incurred the risk of a second attack; the same opinion was expressed by Willan. Having the opportunity of ascertaining whether this lady, who is a mother, and I believe now a grandmother, had ever suffered when her children did, I had inquiry made of her mother. The old lady sent me word that her daughter never had but the one attack; that she had some recollection of my pointing out the crescent shape of the eruption. In this case, slight as the attack was — for there was, as I have said, no catarrhal fever, — it was sufficient to protect the subject of it from any fresh attack. I question if many would be found at the present day who would doubt the protecting power of an attack characterized by the appearance of a true eruption, though all catarrh were absent.

As the rash comes out there may at first be some subsidence of fever, though an increase is possible. Whether this brings an increase or not with the increase of the disease, an increase of fever is to be expected. Often the cough becomes almost incessant and the respiration wheezy. There is a great tendency to a complication with bronchitis or pneumonia. On looking into the mouth patches similar to those on the skin may be seen on the uvula, palate, and tonsils. Though they appear early they remain while the eruption continues on any part of the body. The mucous membranes seem thus to participate in the external manifestation of the disease.

The accession of bronchitis or pneumonia constitutes the chief source of danger in the measles that we are accustomed to meet in the present day. When pneumonia supervenes, the respirations become hurried, the breathing distressed. And if this goes on the expectoration diminishes, the lungs become congested, and less and less pervious. Drowsiness appears, cold feet, in some cases convulsions, and if not relieved, death. There may be red or grey hepatization, but these results that follow ordinary pneumonia are said to be rare: serous exudation through the tissue is more common. Albuminuria has been observed. That deaths from measles are not uncommon the weekly report of mortality in the metropolis shows, but that a great number of these might, humanly speaking, be prevented, I think likely. There can be little doubt but that the children of the poor, often exposed to cold and wet during the primary catarrhal stage, lay the foundation of pneumonia, the peculiar character of which is determined by the disease. It is more the pneumonia of measles than true pneumonia; but still, though bronchial and pneumonic complication may arise even with the utmost care, attacks may be precipitated or intensified by as exciting cause like cold, a very common exciting cause in our variable climate.

Writers on this disease have divided their subject into the milder or common form of measles, and the graver, the rubeola maligna, or morbilli nigri.

Malignant Measles. — Epidemics of this severe form have been from time to time witnessed in this country. When such does occur, either as an epidemic or in isolated cases, the catarrhal fever is more severe, as is also the eruptive stage, which is marked by irregularity; the rash is late in appearing, or it recedes; or it appears early and then recedes; it is dark and livid; there are petechiae or ecchymosed patches. The fever is of a typhoid character. There is long-continued vomiting, tenderness of abdomen, diarrhoea with dark offensive or bloody stools. Delirium and coma are present; the pulse is small; the fauces are of a dark colour. There may be croup, or oedema of the lungs or glottis. The lungs become congested and loaded with muco-serous discharge, and if not relieved death results from exhaustion, from the diarrhoea, or from asphyxia or coma; in some cases from convulsions. As many medical men may be a long time without meeting this formidable array of symptoms, they come to regard measles more as a troublesome than dangerous complaint. They should, however, not forget what may arise, and be prepared to act promptly if it does. Cancrum oris is sometimes witnessed in connection with measles. like scarlet fever measles are very apt to leave dregs. Every organ that has participated may suffer in this way. Besides the complications affecting the air passages already spoken of, phthisis may supervene.

Glandular enlargements, disease of the mesenteric glands may set in; there may be abscesses or ophthalmia.

It was once my lot to witness gout following measles. The patient was an officer, who had gone through the disease favourably. Being anxious to see a friend before he ought to have ventured out, he left home without leave, and was rewarded for his folly with a pretty smart attack of gout.

Measles spreads by infection, and a patient is very rarely the subject of a second attack.

Cases are said to have arisen from other causes than known infection. Thus in the *British Journal of Homeopathy* vol. xxiii, page 263, Dr. Kennedy mentions the case of a boy suddenly attacked, in whose face some mouldy flax seed was thrown. In the twenty-first volume of the same journal Dr. Salisbury, of Newark, Ohio, mentions cases similar to measles produced by sleeping on mouldy straw. He found a fungus growth on the straw. Dr. Kennedy found the same on the flax seed.

I saw a little boy who, while playing with the skin of a hare, suddenly felt itching of his eyelids, rubbed them, and soon had an eruption out very similar to measles. My impression is that none of these cases were true measles, but the possibility of such an occurrence should not be forgotten. These attacks arose, I believe, from Urticaria.

Treatment. — We must now consider the treatment of this disease, and this is, I presume, the most important part of these lectures, though I do not think we could fairly discuss it without refreshing our memory with an outline of the disease we had to deal with.

By common consent *Aconite* is the medicine most frequently ordered in the commencement of measles, as indeed it is in most febrile diseases, and as it is strictly homoeopathic to the symptoms usually met with, good may be looked for from it. If it mitigates the fever (cut it short it assuredly will not) it may lessen the severity of the subsequent stages. It meets the usual febrile symptoms as well as the catarrhal ones, is of use for the feverish restlessness at night as well as for the hoarse hacking cough. Throughout the continuance of the disease we may recur to this medicine to allay the fever at night and relieve the burning heat.

Valuable as *Aconite* may be, there are other medicines that meet the catarrhal stage, and by giving one of them in the day while *Aconite* is reserved for night, more good may arise than from depending on the one alone, though should the accession of fever at night be moderate, or the one medicine meet all the symptoms, there will be no need to have recourse to the *Aconite*.

For the catarrhal stage, when the eyes are very sore, where there is profuse fluent coryza, *Euphrasia* is well indicated. Dr. Richard Hughes has employed this medicine, and Dr. Drysdale, of Liverpool, says that he has been in the habit of using it for years with speedy good effect in the catarrhal stage. It is in this early stage that I would recommend it; and if we remember the importance of mitigating this stage we will see the advantage of using a medicine that will, by helping early, lessen what is to come after. As cough in the daytime is one of the characteristics of this medicine, its use may be greatly helped by *Aconite* at night, which we give to relieve the night cough and allay fever.

Another medicine deserving of attention is *Kali Hydriodicum.* It meets the catarrhal inflammation of the nose and eyelids; it also has among its symptoms profuse papulous eruption on the face and shoulders. It is true that this eruption may be more associated in the minds of medical men with more chronic forms of eruption, but I well recollect many years ago the effect of a fall dose of *Kali Hydriodicum* when the lower extremities were covered very speedily with an eruption very similar to measles in' the shape and colour of the spots; therefore I should, when properly indicated, recommend the use of the Kali up to the time of the coming out of eruption, and if the case progressed favourably through the eruptive fever. Besides these symptoms *Kali Hydriodicum* meets the irritative cough and oppression of breathing, and when the chest is much engaged, may have an advantage over Euphrasia. It seems to be a medicine specially indicated for all symptoms like influenza, which the early stage of measles somewhat resembles.

I would here remark that the *Kali Hydriodicum, Kali Bichromicum, Kali Nitricum, Kali Bromatum,* and the *Kali Carbonicum* are suitable for varieties of cough. Without going into other distinctive symptoms, I would say that the one I have recommended is most suitable for measles, though in some cases where there is hoarse scraping of chest *Kali Bichromicum* maybe of use, but the rash of *Kali Bichromicum* and *Nitricum* is more allied to that of small-pox than measles. A study of the distinctive features of these preparations of *Kali* will amply repay you, as it may enable you to make your selection in some case of difficulty with far greater ease than* if you are not so familiar with them.

The medicine most commonly connected with measles is *Pulsatilla.* Many mothers will venture to treat measles with this medicine and *Aconite,* because they see it recommended in books, and no doubt, with care in the way of nursing, a large number of such cases will do very well; but if pneumonia or bronchitis should set in, they may have cause to regret trusting to their own skill. Now when we look over *Pulsatilla,* we find some of the symptoms of the disease well marked. The redness of the eyelids, the puffed red face, the catarrhal husky cough, the sneezing, and the red spots on the body justify the giving a very prominent place to this medicine. I think, however, that it should not be used indiscriminately as it is. Boehr — who, however, undervalues it — is at a loss to know why it is used. Where there is dry coryza it is much more indicated than where it is moist, when the preference should be given to *Euphrasia* and *Kali Hydriodicum.* If these medicines are given early, *Pulsatilla* will follow well in the eruptive stage; it is also very valuable for the sequelae. It may also help us where the eruption is suppressed or is coming out irregularly. *Pulsatilla* is a useful medicine in the irritability of children, and as we Often see a good deal of this fretting and irritation in connection with the early stage of this exanthem, it is an additional reason for using it.

If the nights are restless and disturbed, wakefulness being a marked symptom, *Coffea* may help. It meets to some extent the coryza; it is also of use for an itching eruption; therefore, though not strictly speaking a medicine that we would rely on too much, for a few special symptoms it may be of use; and, if sleep can be procured, the value of a good night's rest may make a great difference to a patient.

For special symptoms likely to arise we must considerably extend our number of remedies; but it is always well to remember that in choosing for one urgent symptom we should, if possible, get a medicine that meets the case generally.

For a miliary eruption, which would have to be considered along with the other symptoms present, *Aconite, Arsenicum, Belladonna, Bryonia, Ipecacuanha, Mercurius,* or *Rhus* may be given.

Now if we have delirium we will find in *Belladonna* a medicine that will give general help, and we are not overlooking everything else merely to palliate the delirium, for it meets the croupy cough and other symptoms. If we prescribe for one symptom a few doses of the special medicine may be given, but we should return as soon as possible to the medicine for the disease. Thus for a short dry night cough *Hyoscyamus* may at times give help, or *Nux Vomica* for a dry cough; but if we can effect the same good by *Pulsatilla* we should give it the preference, as it meets the greater number of symptoms.

If a case is not going well from the bad coming out of the eruption, from its retrocession, or for its unhealthy character as evinced by the livid dusky appearance or the presence of petechiae, we must be prepared to deal promptly with it. *Pulsatilla* I have already named as a medicine likely to be of use, but the existing symptoms must guide us. Now *Bryonia* covers a good many of these. In the graver forms of measles the fever is apt to become typhoid; *Bryonia* meets this. It is of use also for the cough and for the early stage of pneumonia, also for the petechiae, for epistaxis (for which also *Belladonna* is of use). It is also recommended for the retrocession of the eruption by Hartmann. Use in disease has probably secured this character for it more than its provings; but if in treating one set of symptoms we find another that we were obliged to overlook, (as it did not appear in the provings) yield, and if this occurs at different times, then the medicine may be safely given for this as one of its regular symptoms, if the others correspond.

Ipecacuanha is our chief medicine for vomiting; it is also of use for suppression of eruption, and for teasing cough from tickling in throat, also for rattling noise in breathing. If the eruption is suppressed without oppression of the brain, *Ipecacuanha* may be used; if it fails to do good, or if there are head symptoms. *Opium* must be looked to.

Opium is a medicine that should never be lost sight of where there is a suppressed eruption. If this is the case» if there is drowsiness, or coma> I know no medicine on which I would place greater reliance. It is homoeopathic to the symptoms which we may look for with a suppressed eruption, and it is a medicine that seems to rouse the whole system, and it also, perhaps owing to this, will often greatly help the action of the medicine that we give after it. It should be given every half-hour or hour according to urgency.

When the eruption assumes a dusky hue, when there is blueness or coldness of the skin, petechias, or other symptoms indicating great depression of the system, *Arsenicum* may be given. It may be desirable to give three or four doses of *Opium* rapidly, either before the *Arsenicum* or after a few doses of that medicine, if it does not appear to be acting as it ought. *Bryonia, Arsenicum, Rhus Toxicodendron,* and perhaps *Baptisia,* are medicines to look to in fever of a typhoid character.

In a low putrid state *Secale* may be used as an auxiliary. Petechias, ecchymosis, difficult speech, gangrene, slow, small, intermittent pulse, convulsions, diarrhoea, are indications for it, and under these extreme circumstances we may be very glad of a remedy allied to *Arsenic* on which we can fall back.

The diarrhoea that we meet with may yield to *Pulsatilla*, but *Veratrum* and *Arsenicum*, or with bloody stools *Mercurius*, or *Mercurius Corrosivus*, are more reliable.

The chest symptoms are those most likely to give trouble; even in comparatively mild cases the cough is very worrying; at times it will be incessant for hours together. The medicines we have already spoken of — *Aconite, Belladonna, Euphrasia, Kali Hydriodicum, Ipecacuanha, Pulsatilla, Hyoscyamus,* and *Bryonia* — all meet this state, but one or two other medicines may help: for instance, if we have a very croupy cough, *Iodium* maybe given in rapid doses for a few hours, while warm fomentations are applied over the larynx. If there is wheezing, or slightly loose croupy cough, *Hepar Sulphuris* may be given instead of *Iodium*. *Spongia* also may help for a dry croupy cough, with burning larynx or trachea.

The complication with bronchitis will be met with these medicines. *Phosphorus* also, which is one of our sheet anchors in pneumonia, is also of use in bronchitis. In pneumonia, after the administration of *Aconite, Bryonia,* and *Belladonna, Phosphorus* will often be found useful This medicine is indicated by the hoarseness, cough with raw scraping feeling in chest, dry hacking cough, muco-sanguineous expectoration, difficult breathing, pain and stitches in chest.

Antimonium Tartaricum is a medicine that is indicated in the catarrhal stage, but having named other medicines it is only confusing to increase the number, unless for very special reasons. As the student of homoeopathy comes to test his remedies in actual practice, he may often find an indication for some medicine that he is led to by his repertory, instead of his standard book on homoeopathic practice; but in books on therapeutics, the authors cannot do more than point out the leading medicines with the reasons for their selection, leaving an abundant field from which much valuable fruit can be gleaned. In the later stages of the disease, where pneumonia or bronchitis is present, *Antimonium Tartaricum* must occupy a prominent place. The difficulty of breathing and oppression of chest, with disturbance of heart's action, and debility, are indications for this medicine; but the rattling breathing, with inability to cough up phlegm, or accumulation of phlegm, are the symptoms that induce me to select this medicine. Under these circumstances I have very great confidence in it.

Sepia is of use for a tight cough, rattling, pains in the chest, and purulent expectoration, but as this last symptom is only likely to arise in protracted cases, this medicine will not be so frequently needed as some others; but as it is very valuable for a tight cough, where *Phosphorus* has failed to give all the help we need for this symptom, we may get what we require from a few doses of *Sepia*. In advanced cases *Cuprum Metallicum* should be remembered. Hurried breathing, blue appearance of the face, with weak small pulse, rattling breathing, haemoptysis, discharge of bloody mucus from nose and mouth, asthmatic breathing, convulsive movements — are the indications for its use, and as these symptoms are those of a severe attack, the having such a medicine to fall back on, will be found very satisfactory to those needing it.

The patient must be supported with food, and when necessary stimulants, and whatever aid can be obtained from poulticing must be sought.

For the otorrhoea that may follow measles, *Pulsatilla, Silicea, Hepar. Mercurius* and *Sulphur* may be used. For ophthalmia, *Mercurius Corrosivus, Hepar, Euphrasia,* and *Sulphur* are the medicines on which most reliance may be placed.

As there is usually much debility after a sharp or protracted attack of illness, besides the benefit to be obtained from change of air, and good nourishing diet, such medicines as *China* and *Phosphoric Acid* may be used. Where there has been any drain, as from diarrhoea, China will be useful. Indeed, should there be any chronic tendency to diarrhoea, both these medicines will help. With growing children also always remember that *Calcarea Carbonica* is an excellent constitutional medicine, imparting vigour and counteracting the tendency to struma. Another medicine that will help pale, sickly-looking children is *Ferrum.* The Ix or 2x trituration of the *Saccharine Carbonate of Iron* is one of our useful preparations that I recommend to be employed.

Roseola

This disease is chiefly troublesome from the possibility of its being mistaken for scarlet fever, or what is a greater danger, from scarlet fever being mistaken for roseola. The danger from this cause is, that the precautions that ought to be taken to guard against the spread of infection ate neglected. Parents readily jump at the idea that a slight eruption is only roseola, but as we know that scarlet fever is sometimes exceedingly mild, the diagnosis should always be cautiously made. Happily this disease is exceedingly mild, and not attended with danger. The fever attending it is slight; the eruption following this consists of red or rose-coloured patches of irregular shape, appearing on different parts of the body. Different varieties of this complaint have been described.

The *Roseola Infantilis* has got its name from the age of those attacked; it appears on children during dentition, also at times from intestinal irritation.

The *Roseola Aestiva* appears in the summer months* and is attended with itching and sometimes pain. It is said to be attended at times with sore throat. I must say where a rash of this kind appears and sore throat is present, I would advise you always to be suspicious of its being scarlet fever. You may of course have the combination without this being present, but it will be safer to state what you apprehend and take the necessary precautions, unless morally certain that you have roseola to deal with; you will get no thanks afterwards if you have called the disease *Roseola* and a case or two of scarlet fever follows in other members of the family. Other varieties are spoken of but I need not detain you with them. Dr. Gregory describes a variety as *roseola exanthemetica* or *variolosa,* and I have already mentioned a case that gave some trouble in the diagnosis. I think this hardly comes under the name of roseola further than that a rash of this character seems to usher in some mild form of variola, and it is well to know the possibility of such an occurrence.

Roseola has at times appeared in the epidemic form. The treatment must be regulated according to the symptoms present;

Aconite will be sufficient in most cases. If there should be headache *Belladonna* may be given. If it arises in connection with teething or disordered stomach, *Chamomilla* will be the best medicine. This medicine will also be of use if it arises from emotional causes. If there is much itching *Rhus Toxicodendron* may be given. If there is nothing more than roseola these medicines will be sufficient.

Rötheln

This disease, which was described under its German name by Dr. Paterson of Leith in 1840, has not yet become very familiar to us in this country. We can readily understand how our ancestors (were it common in their day) would, in consequence, have increased difficulty in separating measles from scarlet fever. That many English physicians have not seen it is very probable. Dr. Paterson was so far more fortunate than bis colleagues, as he had sufficient opportunity from the cases that came under his observation to describe it.

The disease is generally supposed to be a hybrid between measles and scariatina. It is said to be scarlet fever *plus* cough, hoarseness, and lachrymation.

There is necessarily considerable difficulty about its real nature, and some questions arise. Do both diseases exist together, as measles and small-pox are said to do (Gregory on Eruptive Fevers, page 134), or is it a new disease resembling both? Were it only to occur once in a way we might say both diseases existed together, but what is described as Rötheln has appeared in a large number of cases, which makes it look more like a separate disease, though of its being such I have grave doubts. It would be interesting to know whether cases of it have appeared at a time that scarlatina and measles have been epidemic together; also whether those who have gone through it have been protected from measles and scarlatina subsequently.

Observers attending to these and some other points may be enabled to clear up much that is now uncertain.

The disease begins with continued rigors, followed by cough, itching, and redness of eyes, lachrymation and sneezing. In addition to these symptoms of measles sore throat is present. Other symptoms that might be expected, nausea, vomiting, headache, and drowsiness are generally present. Rheumatic pains are also complained of in back and chest. After three or four days a general rash breaks out, chiefly on the trunk, this has at first the appearance of measles, the spots (stigmata) run together, forming patches of irregular shape, more red in the centre. In the more severe cases the patches run more together, and are very dark towards the centre. In mild cases the eruption may not remain out longer than four or five days, but in more severe cases six to ten. The sore throat increases during the period of eruption, hoarseness also increases. There is redness and swelling of throat, great difficulty of swallowing, great accumulation of viscid mucus causing sense of suffocation. In some cases coma and convulsions. The eruption may fade away, or the disease may abate on the appearance of some critical evacuation. Desquamation follows. We are indebted to the careful observation of Dr. Paterson and some German writers for our knowledge of this disease, Canstatt describes the disease as a red spotted exanthema, presenting characters of scarlet fever, measles, urticaria, and erythema. It is described by some under the name of *Rubeola,* which is very confusing, as this word has been very commonly used as a name for measles. As sometimes seen, this hybrid varies a little; thus Raue speaks of *Rubeola Scarlatinosa* which corresponds in the affection of the throat and the consequent dropsical affections to scarlet fever, while the eruption is entirely similar to that of. measles. The *rubeola morbillosa* shows all the catarrhal symptoms of measles while the eruption is entirely similar to that of scarlet fever.

These changes of type are confirmatory of the disease being a hybrid rather than a newly described exanthem. At the same time we may wonder that where such a combination is possible, it is not of more frequent occurrence. This disease is very rarely seen, still as it does occur, we cannot pass it over without notice.

The treatment will necessarily consist of much the same class of remedies that are required for measles and scarlet fever, with the addition of one or two . remedies specially applicable to the disease in question. And here it is well to point out to those to whom homoeopathy is new the very great advantage it possesses in having a law to guide us. Hahnemann, before he ever had the opportunity of seeing the cholera, was able, from the description he had received, to say that such and such remedies would be of use, because they corresponded to the symptoms of the disease, and since then we have had abundant opportunity to test the accuracy of his judgment.

So with a disease that is new to us we can at once select medicines that are capable of producing similar symptoms, and we give them with confidence.

After *Aconite, Belladonna,* and such medicines as may be needed at the commencement of the attack, we select a medicine such as *Kali Bichromicum* to meet the symptoms beginning to show themselves. The catarrhal symptoms, hoarseness, cough , and sore throat, and to some extent the eruptions are covered by this medicine. The increase of sore throat may be met by *Hepar Sulphuris,* and at a later stage *Mercurius Iodatus* may be used to counteract the secretion of vitiated mucus. *Lachesis, Opium* and *Cuprum* may also prove of use; indeed, the medicines indicated in measles and scarlatina should be resorted to when needed. To which subjects it will be well to refer.

Chicken-Pox

Varicella is most commonly known as chicken-pox, but it will happen at times that such words as *swinepox* and *bastard-pox* are used to designate the same disease. It is well to be familiar with popular names, as sometimes a question may be asked that a doctor is. expected to be able to answer off-hand, that may puzzle him far more than some question in anatomy or physiology that would have made him quake at his examination. These popular names pass out of use gradually and are giving way to a more regular, if not a more correct popular nomenclature. It was at one time known scientifically as *Chrystal* or *Variolae Chrystallinae.* like the other exanthems, it was not fully described till a comparatively late period. In Morton's time, 1694 it was popularly known by its still common name of chicken-pox. The name of varicella came into use about 1770.

The disease is a very mild one. Isolated cases are at times sharp, but we have no knowledge of the disease ever having assumed a severe epidemic form in any way approaching that of the other eruptive fevers, small-pox, measles, or scarlatina. The disease is contagious and infectious, and has a period of incubation of from four to seven days. The early fever is usually very slight, sometimes hardly perceptible; in some few cases there may be a little extra amount of disturbance as shown by loss of appetite, restlessness, headache, nausea, and wandering pains. This fever is of short duration, perhaps twenty-four hours: it is followed by an eruption of small reddish spots on the back and scalp. They spread to the face, arms, and other parts of the body. The second day they have assumed the form of vesicles, these being as large as a split pea. They are generally distinct, but may be confluent; they contain a faint yellow or colourless fluid. The third day they are at their height, after which they collapse and dry up, or burst, and form a yellowish or brownish granular scab, which soon falls off. The eruption does not all come out at once; it comes out in crops, some are drying up while &esh ones are forming. There is itching. At times a large scab will leave a small pit, but it will only be an isolated one, and this may possibly be due to the assistance derived from the patient's own fingers. About the fifth day the eruption disappears.

An attempt was made by Willan to distinguish the shape of the vesicles, but any such arrangement is doubtful and of no practical value. There is a question, however, that is of some importance, and that is as to whether the vesicles are ever cupped at the top like small-pox.

I have seen the cupped appearance in an undoubted case of chicken-pox, so that there may be a rare exception to the rule. The other characteristics will guide our diagnosis, but the occurrence of cupping may cause a momentary suspension of our judgment.

Treatment. — As the primary fever is so slight the medical man is not usually sent for till the eruption appears, and then he may be sent for hurriedly to say what the nature of the complaint is. Consequently it is either when the vesicles have formed, or are about to do so, that the question of treatment arises. Should he see the child before this stage a little *Aconite* is all that will be needed.

Pulsatilla is a medicine that may be given if there is fretfulness and irritability with the eruption. Our little patients sometimes supply this characteristic when annoyed by the eruption, or perhaps be being kept in bed when they do not feel inclined to stop there.

Where there is much itching *Mercurius* will be more suitable.

Should the vesicles be large and look angry, or should cough accompany the attack, which often is the case with eruptive fevers, from some extension of the irritation on skin to the mucous surfaces *Antimonium Tartaricum* may be given.

It is always safe to guard against cold, and though the child need not be kept on low diet, care should be taken not too give anything to irritate the bowels and so induce diarrhoea.

As the rash is disappearing a dose of *sulphur* may be given night and morning. Strumous children are apt to suffer after exanthematous diseases, and as this is a pretty wide-spread form of constitution, *sulphur* which helps to prevent the development of scrofulous complaints, and to remove many of their symptoms when present, is a safe medicine to give for three or four days after any of these attacks.

Erysipelas. – (*St. Anthony's Fire*)

This disease, though according to the strict meaning of the word, an exanthem, differs so much from the ordinary eruptive fevers that the propriety of classing it with them is open to question. It appears to arise spontaneously, though it is very doubtful if it really does so. It would seem as if the germs of this disease are floating about in our atmosphere, and only want a suitable soil on which to plant themselves, when they will burst into life and furnish fresh sources of infection. This is a wide field for speculation, and as the evidence that would furnish the proof needed is in our present state of knowledge unattainable, we can only draw inferences from certain broad facts that come before us. Eruptive fevers arise from contagion or infection, erysipelas also spreads in this way, but it will show itself after some slight wound, where no other cause is known to exist It will also show itself in a person of broken-down constitution, and it shows a tendency to return again and again in the same individual It is liable to be conveyed from one person to another by an intermediate agent, as scarlatina is also conveyed; and it has the remarkable property, at least in one instance, of producing a disease different from itself. Thus puerperal fever has been known to take its rise from the infection of erysipelas, and M. Bouchut, speaking of the erysipelas of infants, says, "erysipelas is never more frequent than in the course of the epidemics of puerperal fever. The knowledge of this connection between these diseases ought to make medical men very careful, and to decline attending a confinement where from their attendance on erysipelas there is any possibility of their being the means of propagating such a dire disease as puerperal fever."

Since the above was written I have received a letter from the Rev. A. F. Elsner, a very intelligent and devoted missionary to the Eskimo. He writes from Hopedale, Labrador. Being a German some of his expressions are a little peculiar.

"A boat's company of our people (Eskimo) brought from about forty to fifty miles southward the corpse of a young woman who had died in childbed, and as it afterwards was evident from puerperal fever; some of them had thick swollen faces and severe headache, which together soon proved to be erysipelas, and very infectious; its chief seat was the head, but often also hands and arms. People who had a small sore spot, a scratch of a nail, or the like, were very much apt to get it, and it worked from that spot over a great part of the body, often going over in gangrene and causing death, or in suppuration with very deep ulcers, on some of which we had to cure a quarter of the year. It happened that just at this time seven women were delivered, and all seven got the child-bed fever: four of them died and three recovered.

"In the first sickness (erysipelas) *Belladonna* and *Mercurius* seemed to act pretty well, but often proved insufficient. Most benefit, I think, we saw from *Nitric Acid,* sometimes in connection with other remedies, in order to cover secondary symptoms.

"Misery was then very great in the poor small houses of our Eskimo, and the want of linen and suitable food w:as felt often keenly. We have done our utmost to assist them both ways. Worst of all was being compelled to use the same medicine; our small stock of it was soon exhausted, and even our private homoeopathic medicine boxes partly emptied, so we often only could give a remedy which we thought to be next to the desired one." Their troubles were not at an end, for whooping-cough appeared afterwards probably for the first time in Labrador, but as this is not connected with our subject I need not quote further."

The Moravian missionaries in most of their settlements, but especially in Labrador, are so entirely out of reach of medical aid that they must depend on their own skill. Mr. Elsner has established a reputation for himself as medical adviser in the stations he has been at. Long experience has made him familiar with disease, and his annual letter to me usually contains many matters of interest The mission ship, the *Harmony*, takes out the letters in June, and brings back the missionary intelligence about October. Surely the men that thus labour for years in such a climate ought to excite our admiration, and put some of us at home to shame for our want of zeal.

Erysipelas at times takes on the epidemic form, and operations become very unsafe when this is the case. A remarkable feature in connection with spread of this complaint is the well-known fact, that cases within reach lying in one direction will suffer while others escape, thus one side of a ward, or in a given line across a ward, patients are affected, while adjoining beds escape, as if the epidemic wave was of limited extent.

The period of latency is from two to fourteen days. — *Atkin.*

Erysipelas consists of a certain amount of disturbance of the system, sometimes scarcely perceptible until after the appearance of the eruption, which commences with a red spot and swelling, the redness deepening and extending from the centre. A rosy blush of varying size surrounds the deeper red. This efflorescence travels over the skin, passing from one side of the face to the other, or over the body, and returning at times to the original point of departure or attacking a distant part: as it moves onward the part first attacked loses its redness and swelling, and some desquamation may alone mark what has been the seat of severe inflammation of the skin. Pressure causes pain, a depression and whitish appearance is produced, but this is speedily replaced by the return of the deep red colour. The disease does not confine itself to the surface, the cellular tissue beneath suffers, and suppuration and sloughing — sometimes very extensive— show the severity of the attack.

Vesicles or bullae containing yellowish serum often appear on the skin. Oedematous swelling is also frequently seen.

Though an erysipelatous redness is often the first sign of an attack, yet in other cases it sets in with rigors, nausea, vomiting, or diarrhoea, headache and delirium, the efflorescence appearing some hours, or a day or two later. Erysipelas will also appear at times in the course of another disease; thus, in connection with the later stages of scarlet fever, or during the progress of vaccination. I have never myself seen anything like a severe attack of erysipelas following vaccination, though I have seen a deep erysipelatous redness covering the greater part of the arm from the shoulder to the elbow, attended with much tumefaction, but this has always subsided in a short time. Death has ensued from erysipelas thus induced. I shall, however, have to allude to this again when speaking of vaccination. Such an event must be extremely rare, or we should hear a good deal more of it from the anti-vaccinators. If much fever accompanies erysipelas, it assumes a low type, the state of the patient being similar to that of typhus, death resulting from coma, or debility. When suppuration and sloughing takes place to a great extent, the patient may die from exhaustion; but it is wonderful at times the drain that will take place, and yet recovery follow at a protracted period.

Inflammation of the brain, of the abdominal or thoracic viscera, and oedema of the glottis, are all sources of danger in this disease. Pyaemia also is not an unknot occurrence. Deposit of serum in the cavities of the body, or of sero-purulent fluid, may take place.

There is a form of this disease that occurs in connection with children, known as *infantile erysipelas*. It has been chiefly observed amongst those born in hospitals, and has been extremely fatal. Improved sanitary arrangements have no doubt contributed to the lessening of this disease and consequent mortality; but as it arises amongst badly fed children, and is associated with the prevalence of puerperal fever and other diseases of a malignant type, we are at any time liable to meet with isolated or even numerous cases. Our freedom from a disease for a lengthened period should never allow us to shut our eyes to the possibility of its return.

The neighbourhood of the umbilical cord and lower part of abdomen is the part most commonly attacked in the very young. The limbs, however, are at times attacked. It commences as a dark red, shining, purplish-looking spot, with hardness but hot much swelling. The disease is apt to assume the vesicular form and spread to the cellular tissue: portions of the skin may be destroyed by gangrene, the scrotum has been destroyed in this way, leaving the testicles bare; but terrible as such a case is, recovery has taken place.

The child is restless, continually crying, the face pale, pulse feeble and frequent, vomiting and diarrhoea are often present, convulsions also are apt to supervene. Where the abdomen is the part attacked, serous effusions, adhesions, and plastic exudations will generally be found in those who die. Infiltration into the cellular tissue, and sero-purulent deposits and sloughing where other parts are engaged.

Treatment. — One of our first objects must be to sustain the patient's strength; therefore in treating adults, such light nourishment as can be borne should be given, together with stimulants. Where there is vomiting this must be allayed by *Ipecacuanha* or *Arsenicum,* and beef tea given in small quantities till the stomach can retain it. During recovery, the patient will need plenty of support.

In the case of infants, good milk should be given diluted with water according to age, or asses' milk, if the mother's milk is not of a good quality Maunsell and Evanson speak highly of white wine whey, a teaspoonful every hour or half-hour. Where there is great debility, two or more drops according to the age of the child, of whisky, in a little milk or sugar and water may be given as needed; a small quantity given cautiously helps an infant very much, — but as often and in as large quantities as each case seems to require —

will often help to bring a child through. A small quantity of beef tea in addition to the milk will be a great improvement; it tells quickly on a young infant, but the amount had better not be large. It is as an auxiliary to the milk that I find it of use.

The usual local treatment may be used, dusting the part affected with flour, starch powder, or arrowroot, or when there is suppuration, or sloughing, or a threatening of either, fomentations and light poultices. Incisions are sometimes needed.

One of our chief remedies is *Belladonna;* the redness, swelling, and head symptoms, when they are present, point to this medicine as a very valuable one,

Rhus where there is vesicular erysipelas; it also meets the state of fever which is of the character of typhus.

Apis is a medicine that will be found of great value where there is bright red swelling and shining, also where oedema is present. Where oedema of glottis threatens, this medicine must be our sheet anchor. though *Arsenicum* and *Lachesis* may help under these circumstances.

If the chief characteristics of these medicines are borne in mind:—
Swelling, redness and head symptoms, *Belladonna.*
Vesicular eruption and typhoid state, *Rhus.*
Oedema, with shining swelling as if stung, *Apis.*

Many ordinary cases of erysipelas require no further medicine than the one first chosen, but as cases become more complicated a greater variety will be needed.

Croton Tiglium, a medicine that is, I fear, too much overlooked. In this disease it meets the redness, swelling, vesicles, and oedema. Itching, tendency to gangrene, debility and diarrhoea, when present, as well as feeble pulse and headache. I should expect it would also be of much service in oedema of glottis. Therefore, whether in the erysipelas of adults or of children, this medicine may be used early, as well as in the approach of the more severe stages.

Aconite will be required for the hot fever at night, but if there is much head disturbance and delirium, *Belladonna* will very likely be sufficient.

Arsenicum may help in any severe case, where there is great debility, also for the following — Vesicles; bloody blisters; black, burning and painful blisters; swelling with itching and burning; gangrene; coldness; sleeplessness with uneasy tossing about; anxiety; heat and throbbing in head; dreams; vomiting. In the advanced stages it is likely to be most needed.

Lachesis is a medicine allied to *Arsenicum,* and that may aid if indicated. Thus, gangrenous blisters, swelling of cervical glands, blackness of lips, soreness of throat, despondency, heart complications, headache, swelling of head and face, would lead us to think of it.

Secale also where there is gangrene and great debility, bloody blisters on extremities, petechise, restlessness at night, and convulsive movements.

Nitric Acid, as found of use by my friend Mr. Elsner, may be given where there is erysipelatous inflammation and swelling of cheek, stinging and painful, with nausea and chilliness, putrid smell from mouth, vesicles on tongue. The fever would be met by this medicine; also shuddering, violent fever with chilliness, swelling of submaxillary glands, dysenteric or bloody diarrhoea.

Comocladia Dentata, one of the new American remedies, has under its symptoms, violent itching, redness and erysipelatous swelling of the face, hands, and other parts of the body, followed by yellow vesication and desquamation of cuticle.

Painful burning on face and arms; face enormously swollen.

Inflammation of skin, followed by ulcers discharging purulent foetid matter.

Doryphora Decem-lineata is recommended for erysipelas of face and head and cornea.

Plantago.— Swellings deep redness, itching, restlessness, wounds showing a tendency to sphacitus and suppuration.

Veratrum Viride, which seems to exercise a powerful influence over fevers, inflammatory affections and congestions, and to act on the cerebro-spinal system. Has been recommended in erysipelas, phlegmonous or vesicular, applied topically as a lotion in the proportion of 3i to 3iv. — and internally in the 1x dilution. I have no experience of this myself, but give it on the authority of *Hale's New Remedies.*

It is impossible to go through all the indications that may call for any of these medicines, all that can be done is to point out some of their leading features where they are likely to be useful. The repertory must be studied for more careful grouping of symptoms.

Acid. Phosphoricum. Where debility, especially attended with a weakened state of nervous system, remains, should be looked to.

Hepar should be used where there is suppuration. *China* may be given where there is debility following long illness, or exhaustion from loss of fluids.

Silicea where the suppuration is of a chronic and exhaustive character.

Phosphorus for chronic, slowly suppurating abscesses.

Small-Pox

Variola. — This disease was described by Rhazes about the year 910. The descriptions of epidemics earlier than this date leave us in doubt as to whether small-pox is the disease alluded to or not. Procopius gives an account of an outbreak of disease in Egypt, in the year 544 that some think was small-pox; but where we can only conjecture as to what may have been, as the evidence is incomplete, we must rest content with the knowledge that prior to the time of Rhazes the early history of this malady is hidden in obscurity. The name Variola is derived from *varus,* a spot or pimple. The term *pock* signifies a vesicle or bladder, a word of Saxon origin. The addition of *small* is of comparatively late date, and appears to have got into popular use to show the form meant.

As already mentioned, the great Sydenham pointed out the difference between small-pox and measles. This writer makes an observation that sounds strange to us at the present day. Speaking of epileptic fits in connection with small-pox, he says, "When these seize infants who have just done teething, I always suspect that small-pox is on the road, and generally its eruption verifies and confirms my prophecy by breaking out within a very few hours, so closely, that if an infant, as is generally the case, has a fit of convulsions in the evening of one day, small-pox will show itself by the morning of the next. I have also noticed that the small-pox which attacks infants immediately after fits of this sort presents full-sized pustules, which are rarely confluent, the disease being mild and of a good sort. The tendency, however, to perspiration I have never observed in infants either before the eruption or after; it seems limited to adults." But for the change made by vaccination I might now either be speaking approvingly of these views, or the reverse. This admirable physician carefully noted the progress of the disease, and gives some valuable advice about treatment, warning against over-heating, and thinking of the comfort of the patient in allowing him to change his position in bed; thus raising his voice against some of the absurdities and barbarities that prevailed before his time. He also gives a sly hit at some of the habits of the people in his day that prevail still amongst us. He says, "Now-a-days every house has its old woman, a practitioner in an art which she never learnt, to the killing of mankind."

Now, when we contrast the wisdom of Sydenham with the barbarism of Diemerbroeck, we cannot but see what a blessing it was that the powers God had bestowed on this man should have been used for the advancement of medicine. Diemerbroeck says, "Never shift the patient's linen till after the fourteenth day, for fear of striking in the pock, to the irrecoverable ruin of the patient. Far better is it to let the patient bear with the stench than to let him change his linen, and thus be the cause of his own death. Nevertheless, if a change be absolutely necessary, be sure that he puts on the foul linen that he put off before he fell sick, and above all things take care that this supply of semi-clean linen be well warmed." We may wonder that any of this man's patients recovered. If they did so, we can only say they must have had fine constitutions.

Small-pox spreads by contagion and infection. Could all sources of these be destroyed the disease would die out. Unhappily these sources are multiplied; thus, according to Humboldt, the disease was introduced into Mexico by a slave in 1520. Of its terrible ravages among the coloured races history tells us. That an unsuspected source of contagion or infection may exist, and break into activity from some disturbing cause, I have no doubt. I recollect being in the old small-pox hospital the last day it was occupied by patients. This building stood where the Great Northern Railway terminus at King's Cross now is. After the house was given over to those who buy old houses, it was pulled down, the materials would in the ordinary course of trade be sold, and dispersed about London, to be again used in building. The year following the pulling down of the hospital we had an outbreak of small-pox in London.

We are constantly exposed to danger from the reckless conduct of others, who forget how true it is —
"Evil is wrought by want of thought
As well as want of heart."

It is but recently that it has been made a punishable offence to remove persons in public conveyances suffering from infectious diseases. How many lives have been lost from disease spread in this way! I some time since noticed that in the breaking up of an old street sewer, the bricks were trimmed and made ready for use before removal My attention being aroused, I asked a man who dealt in old houses, what could be done with these. He told me they would probably be used for making party walls!!! A pleasant idea that we and some unsuspecting neighbour might some day share the risk of having these filthy bricks separating our houses, I felt it my duty to write to one of the medical officers of health calling his attention to the matter, but doubt if he had any power to interfere.

I have seen part of an old bedstead that had been utilized in forming part of the roof of a handsome West End house, showing that our ancestors knew something of what is called "scamping" as well as our modern workmen.

I was with Dr. Gregory one day when he received a letter from a medical man in Kent, telling him that a graveyard had been opened in some country district, in which the victims of an epidemic of small-pox twenty years before, had been buried, and that the opening of the graves had been followed by an outbreak of small-pox, and asking him if he thought the cases now appearing had been caused by this. He thought not — that it was merely a coincidence. Neither did he think the distribution of the materials of the old small-pox hospital had anything to do with the cases that appeared in London after it. Much as I valued the opinion of my friend in many things, I could not but think that in these instances he was wrong. It is the more strange, for he says in his book, "The dry scabs of small-pox retain a contagious property for a great length of time. Experience, too, has taught us that death does not destroy the energy of the purulent secretion. A child has been successfully inoculated with matter taken from the dead body;" he does not say who performed this daring experiment, but it is one that I trust will find but few imitators. He goes on to say, "A confluent case will taint the air, and spread infection for at least ten or twelve days after death. The knowledge of this fact has induced the Secretary of State to issue orders that the bodies of those dying of small-pox are not to be admitted into the schools of anatomy." (Gregory, on "Eruptive Fevers.") Our author, while recognising a long continuance of infection, would seem to have thought, from what he said to me, that after a given lapse of time this power would be lost.

It is these hidden and dormant sources of infection that I believe produce many of those cases of disease that appear to arise sporadically. So far as we are aware, the disease never breaks out in places that it has not visited before; thus it was unknown in America till we introduced it. The legitimate conclusion is, that, however the disease was first introduced into our world, it now never arises unless from some source of infection. like the other exanthematous diseases, it has frequently taken on the epidemic form, and has committed great ravages. It is but recently that very general alarm existed on account of the large number of deaths from smallpox, and now we sometimes are gratified by reading in the weekly return of deaths in the metropolis, "Deaths by small-pox, none." [1] I need not delay you with any history of these epidemics, nor will I say anything on the important revolution that has been effected by the introduction, first of inoculation, and then of vaccination, until after speaking of the disease itself and its treatment; not omitting to mention that variety known as "modified smallpox," the peculiar character of which is due to vaccination, though before its introduction this mild form was occasionally met with.

Small-pox has a period of incubation of from about ten to sixteen days, twelve being the most common. When it results from inoculation this period is from seven to nine. The period of latency is terminated by the accession of rigors. There may be a rigor which is followed by burning heat, which is interrupted by the accession of another rigor. With these symptoms there is severe headache, muscular pains, and weakness. There is weakness in back, also pain in back and loins, which is a very characteristic symptom. When this is present with other febrile symptoms, if the disease is at all prevalent, it should at once arouse suspicion. Tenderness on pressure in the epigastrium is complained of£ The face is flushed, there is increased beating of the carotid and temporal arteries; the tongue is furred, nausea and vomiting are also present. At times there is syncope; the pulse is quick and full, the skin hot and dry, but sweating is often a marked symptom in distinct small-pox, sometimes preceding the eruption for a short time, in other cases for a longer period. Trosseau lays much stress on sweating as a characteristic symptom; where sweating is absent, which is apt to be the case in confluent smallpox, there may be a hot dry skin. Delirium and restlessness may be looked for, in some cases drowsiness and coma are present; and convulsions, especially in the case of infants, are not uncommon. The urine is diminished and high-coloured, and bowels often confined. These various symptoms, in more or less severity, are followed by the appearance of an eruption, which may be looked for about forty-eight hours after the commencement of the rigor ushering in the fever. So that fourteen days after exposure to infection the rash may be appearing. As this comes out, the vomiting which is often frequent and distressing abates, and the febrile symptoms, including the headache, diminish, as well as the pain in back. The temperature on the first or second day of the primary fever reaches to 104 or even 105. "During the initial stage," says Wunderlich, "the temperature affords no aid in deciding on the severity of the disease, but the course after the eruption aids us very much. Complications, also, for the most part may be recognised by the temperature when occurring after the first commencement of the eruption." When the maximum has been reached, a slight fall immediately ensues, which generally lasts one day. At this time we may commonly notice the first traces of the eruption in the form of spots. This stage lasts from two to five days, and it is not possible at this time from the course of the temperature to discriminate smallpox from exanthematic typhus, relapsing fever, or from a pneumonia which as yet affords no local evidence of its presence." He further says, "If the fifth day of the disease pass without any eruption making its appearance, the presence of small-pox must be considered very doubtful.

As the eruption becomes developed the febrile symptoms abate, and with them the temperature falls more or less rapidly; and if a normal temperature is quickly reached, modified small-pox is the form of disease we may look for. In other cases, as the eruption on skin becomes more fully developed, the temperature again rises; if the case is not very severe the temperature may range from a little over 102F to 104. As suppuration goes on the increase of temperature above 104 will indicate danger. Death may take place at a moderate temperature, but in fatal cases 107.6 may be reached, or higher; and Simon, as quoted by Wunderlich, has measured a temperature of 112.1 after death.

The chief point to remember is that a high temperature, after the appearance of the eruption, indicates a more severe attack or the presence of complications,

As the eruption comes out, anxiety is felt to know what form of the disease is about to show itself.

The varieties are. *Variola Discreta,* or distinct small-pox; *Variola Confluens,* confluent smallpox; *Variola Haemorrhagica,* black small-pox; and *Varioloid,* or modified small-pox. Of these we have now to speak, and also of the stages of the disease. There have been more minute divisions, but for all practical purposes these are sufficient.

Discrete small-pox, for such is the quaint old word that has come down to us to describe the distinct variety, shows itself forty-eight hours after the rigors, this period marking the first or febrile stage. Owing to debility the eruption may be delayed to seventy-two hours. It comes out first on the face, forehead, and wrists. Sometimes it has been observed to commence on the lower extremities, and sometimes a few large papillae precede the general eruption, and become vesicles before the surface is extensively occupied. An inspection of the velum palati may show an early appearance of the eruption. On passing the finger over the papules they will be observed to be distinctly raised. The *third* or *fourth* day the eruption extends itself over the face, and gradually spreads over the body, being abundant on the parts that are exposed.

The eruption comes out in three crops, — the first on the face, neck and arms, the second on the trunk, and the third on the legs. Several hours intervene between each crop. As it is beginning to disappear on the face, it is ripening on the lower extremities. Regularity in this disease is looked upon favourably, it being considered a sign of a strong constitution. Writers seem to lay even more stress on this than on the severity of the disease. And no doubt, where the vital powers are good, and can be well sustained, a great amount of disease can be endured before the patient succumbs.

It is, however, difficult to predict early in the disease what a later stage may produce. Thus I lately saw a case of confluent small-pox, in consultation with Dr. Hastings, of Brixton, where a fair amount of strength, and some abatement in the delirium, gave reason to hope for a favourable result. The process of Maturation had however to be gone through, and during this the young lady's strength gave way. This stage of the disease is necessarily a trying one, from the increase of surface mischief and the continuance of fever, when the patient is weakened by what has gone before.

About the *fifth* day the papule has been converted into a vesicle containing a colourless fluid, and having a central depression and an inflamed areola. The *sixth* day other symptoms appear. Hoarseness, dysphagia, swelling of throat; the lining membrane of the mouth is affected, viscid saliva is secreted, the eyelids become swollen, there may be swelling of labia or prepuce. The swelling of the eyelids may cause temporary blindness.

The *vesicular* stage lasts about four days, when the vesicle is converted into a pustule; this is said to ripen. About the *fifth* or *sixth* day the fluid contents of the vesicle may be observed to be changing colour. Sir Thomas Watson points out that while the vesicle still exists there is one portion pustular, containing pus, that can be punctured without letting the contents of the vesicular portion escape.

The *pustular* stage lasts about three days. About the *eighth* day, when the pustules have attained their full size, or have ripened, or *maturated,* a brown central spot may be observed on the top of each; a rupture takes place here, the contents exude and form a scab, and the pustule shrivels up. The scabbing lasts about three days, so that the whole process is completed in about eleven days, or if delayed, the scab may not be off till the fourteenth day. The pustules may be at their maximum on the trunk the eleventh day, but continue to increase on the hands till the fourteenth. The place occupied by a pustule is marked by reddish brown discolorations, which become of a dead white colour if the *rete mucosum* has been injured by ulceration, resulting in the well-known pit of small-pox. In mild cases, and on the lower extremities, the vesicles may shrink prematurely, forming only a scabby crust.

Speaking of the pock, Rindfleisch says, in his "Pathological Histology," page 349, vol. i., "The pock originates as a papule on a very hypersemic base." This proposition, which is repeated by all textbooks, may be allowed to pass unchallenged if we choose to term every hard nodular elevation of the cutaneous surface a papule. But it must be remembered that the small-pox papule differs very essentially from the papules we have already described (*e.g.* the papule of measles). The smallpox papule is situated, at least in great part, in the epidermis, not underneath it. It begins apart from the hyperaemic condition of the papillary body, as a circumscribed "parenchymatous inflammation" of the epidermis. He then says, "I use this word under reserve."

He says the middle stratum of the epidermis is the starting-point, which he describes as composed of "intermediate cells," and as belonging to the mucous layer.

It is convenient to consider "the skin as made up of the epidermis and papillary body on the one hand, of the corium and subcutaneous areolar tissue on the other." "The papillary body, apart from its other physiological functions, is the matrix of the epidermis."

According to Trosseau's observation, adults suffer most from constipation, children from diarrhoea. Children sleep more, and the tendency to convulsions noticed in them, so long ago as Sydenham's time, is important; they are observed in some cases to squint, and grind their teeth.

The pain in the loins may be attended with numbness in lower extremities and retention of urine.

The fever that has subsided with the appearance of the eruption again comes on during its progress. Its amount depends on the severity of the case. This accession of fever may be looked for the *seventh* or *eighth* day. It is the fever of suppuration, which is perhaps more correct than the word secondary fever, which is sometimes used. As the scabbing goes on the fever abates, and the temperature returns to its natural standard, provided the case is free from complications. The possibility of internal mischief, or of inflammation of the ovary or testicles, must not be forgotten.

Tenderness of the skin is observed during the process of maturation, more in delicate-skinned women than in men: it is not considered an unfavourable sign. The itching is often considerable. Another circumstance connected with small-pox is the peculiar unpleasant smell given off by the variolous matter.

During the last epidemic a gentleman, who had been a missionary in India, was stopping at a house on Haverstock Hill; being a stranger, a servant was sent with him to show him the way to the Hampstead railway station. As they were walking along he began to sniff with his nose, and said, "You have small-pox here." They were passing at the time the temporary small-pox hospital, the existence of which he was ignorant of.

Variola Confluens. — This variety derives its name from the running together of the pustules.

The fever preceding this form of the disease is more intense, with less sign of abatement on the appearance of eruption; the temperature may fall slightly, but the fever continues. Coma, delirium, and diarrhoea are more frequent.

This latter is observed even in adults. The eruption appears earlier. As already noticed, Trousseau lays a good deal of stress on the presence of perspiration in distinct small-pox, and its absence or lessened quantity in confluent. In confluent and semi-confluent small-pox the eruption is more disposed to spread into mouth, nose, and trachea, giving rise to increased difficulty of swallowing, hoarseness, and cough. This is at first dry, but ultimately expectoration and a general copious viscid secretion comes on. Salivation is a common symptom. The salivation is sometimes excessive, causing great distress to the patient. Glossitis may be present. Oedema about the larynx, and thickened membrane, interfere considerably with respiration, causing dyspnoea. The face, hands, and feet are swollen. The tongue swells and is purple in severe cases. The swelling of the face increases to about end of *ninth* day, and begins to decrease on the *eleventh*. There is often, but not always, considerable cellular complication; the scalp is often swollen; at times confluence without much swelling is met with. Abscesses may form, or there may be sloughing. The rash does not come out with the regularity observed in distinct small-pox. There is usually a redness of the skin, which is swollen, preceding the eruption. About the second day some red points appear, and the third day the vesicles begin to appear and become confluent; they maturate earlier. They are not acuminated, but are flat and irregular, and instead of pus contain a brownish ichor. The inflammation of skin at times is of an erysipelatous character; blebs form in these cases, a thin ichor exuding from them; large bullae as big as a hen's egg are sometimes formed. The swelling is greatest about the sixth day, the salivation about the ninth or tenth, the swelling going down before the disappearance of the salivation, which lasts till the eleventh day.

Sudden subsidence of swelling, or absence of swelling and redness, in the intermediate spaces, as also the cessation of salivation, or its absence, are regarded as unfavourable symptoms.

In some cases the salivation ceases, then returns; it is less common in children than adults, but they may have dangerous diarrhoea in its place.

The natural abatement of the swelling of face, as occurs in *discrete* small-pox, with the breaking of the pustules about the ninth day, must not be mistaken for a symptom of danger.

The exacerbation that takes place in confluent small-pox about the ninth day sounds a sufficient note of warning as to the increase of danger.

About the ninth day of the disease there is often considerable exacerbation of fever, the tongue becoming browner, the eruption becoming black or livid, while petechiae and other symptoms of typhoid or purpura, such as haemorrhages, show themselves; cough with haemoptysis, or haematuria, are not uncommon, dysentery also comes on in some cases.

The appearance of a black spot in the centre of the pock, or a disposition to gangrene, is an unfavourable symptom. The salivary glands partake of the inflammation. Glossitis also, and swelling of the neck, and possibly oedema of the glottis, may increase the danger and account for some of the suddenly fatal cases. Increase of delirium, or coma and convulsions, may be looked for, and, as might be expected from the time the exacerbation comes on, the eleventh is often a fatal day in this disease.

In cases that recover the pitting is much more extensive than after distinct small-pox, owing to the greater extent of skin affected, as it is difficult sometimes to find a sound piece. When we remember the shock the system receives" from a burn, even of apparently moderate size, we must not be surprised at the fatal issue of many of these cases, considering the large surface of skin that is so sharply affected.

Sydenham considered the fourteenth and seventeenth days as periods of anxiety if the patient got over the immediate danger from the suppurative fever. When pregnancy exists, miscarriage usually takes place the seventh or eighth day, as the pustules suppurate, and death a couple of days later. The disease is very fatal to pregnant women.

After such a terrible illness, if recovery takes place, convalescence is naturally protracted, and serious mischief may retard it, or permanent evil remain. Thus abscesses, lameness from inflammation of a joint, otitis, ulceration of cornea, and blindness are dangers to be apprehended; the disfigurement, from the scars on the face, however, being if possible even more dreaded.

The following remarks of Mr. Marson, who for many years most ably filled the post of resident surgeon to the small-pox hospital, are quoted by Dr. Aitken, and state the case so clearly that it is well to repeat them. "The eye seems to possess a complete immunity from the small-pox eruption, and although it sometimes extends to the inner margins of the eyelids, the particular local affection that causes the destruction of the organ of vision in variola begins generally on the eleventh or twelfth day, or later, from the first appearance of the eruption, and when the pustules in every other part -of the body are subsiding. It comes on after the secondary fever has commenced, with redness and slight pain in the part affected, and very soon an ulcer is formed, having its seat almost invariably at the margin of the cornea. This continues to spread with more or less rapidity, and the ulceration passes through the different layers of the cornea, until the aqueous humour escapes, or till the iris protrudes. In the worst cases there is usually hypopium, and when the. matter is discharged, the crystalline lens and vitreous humour escape. In some instances the ulceration proceeds very rapidly: I have more than once seen the entire cornea swept away within forty eight hours from the apparent commencement of the ulceration; and what is singular, now and then the mischief goes on without the least pain to the patient, or his being aware that anything is amiss with his eyes." Further, he calculates that in 1,000 cases twenty-six had ophthalmia, or about one in thirty-nine; and of those, eleven lost an eye each, or one in about 100.

Happily, since the more zealous prosecution of vaccination we see nothing like the number of blind or disfigured people that we did some thirty years ago.

The following remarks of Dr. Gregory are important: — "Nurses will talk to you of a five, six, seven, eight, nine, and even ten days' pock. They are quite right. When the disorder is perfectly normal in its course, not interfered with by any peculiarity of habit, either congenital or acquired by previous vaccination, — when the constitution is sound, with sufficient strength of system, and a good but not over-abundant supply of blood, — lastly, when there is no*t too copious a crop on the surface, the pock maturates in seven days. In severe cases of a semi-confluent or corymbose kind, the process of maturation occupies eight days, — in bad confluent cases nine, or perhaps ten. On the other hand, after vaccination, or when there is some originally favourable diathesis present, the pock will maturate in six, or sometimes imperfectly in five days. This five-day pock constitutes the mild, mitigated, or modified form of variola, now so familiar to us as occurring in those who in early life had been well vaccinated. But this variety of the disease, though formerly less frequent, was yet well known to all the old authors. Van Swieten describes it under the title of variola verrucosa and cornea (stone pock, horn pock, and wart pock).

Variola Haemorrhagica or black small-pox, the variolae nigrae of the old writers, was more common formerly. It has appeared more frequently with the irregularly developed eruption than with the confluent variety. The skin is more or less marked by red petechiae, varying in size from a pin's head to a quarter of an inch in diameter. Most frequently the disease appears as if about to be confluent, but nature lacks the power to throw it out; haemorrhage takes place simultaneously with the eruption of the papules on the third or fourth day, from the free surface of the body. Haemorrhage may come on from nose, mouth, intestines, or vagina. There is ecchymosis of conjunctivae; the petechiae become dark purple, and death takes place suddenly and easily from the fourth to the sixth day.

Congestions, blood clots, and ecchymosis are found after death, the left side of heart empty, and congestion and clots on the right side.

Varioloid, or modified small-pox.

This form of the disease, though not unknown to the older physicians, as we have already seen, has become very common since the introduction of vaccination. It is ushered in with a great deal of fever. There is at times intense headache, great prostration, and vomiting. The eruptive fever is of shorter duration than in the other forms, one day or less, though it will in some cases extend to thirty-six hours, or perhaps more. The eruption comes -out first on wrists and face, is usually distinct, but shows a good deal of variety; some pimples may be abortive, some of the vesicles shrivel up (horn pock); when unusually persistent and transparent (pearl pock), some of the pustules show the characteristic umbilicated or central depression. Unless some complication takes place, or from the fulness of the eruption there is some fever attending it. the worst part of the attack is over as the eruption comes out.

Cases of small-pox without eruption have been observed, giving immunity, as the disease has not been taken a second time when the subject of it has been exposed to infection. Some fatal cases have also been observed. It is difficult to account for this modification, as such cases occurred before vaccination, the primary fever indicating the nature of the disease.

Mortality in Small-pox. — In a recent epidemic of small-pox in Chili, the disease became very confluent, and haemorrhagic, the mortality for a short period attained to fifty per cent. These ravages occurred chiefly among the indigent and non-vaccinated. — *Lond, Med. Record,* Dec. 1876.

This disease has been very fatal with children. Dr. Marson has pointed out that it is most fatal with children and old people. According to this authority patients of all ages die at the rate of fifty per cent, in confluent small-pox, at the rate of eight per cent, in semi-confluent, and at the rate of four per cent, in distinct small-pox.

Diagnosis of Small-pox. — Before the appearance of the eruption small-pox may readily be mistaken for other diseases. The pains and sweating may lead to a suspicion of rheumatic fever. The lumbar pain and vomiting with high fever may point to relapsing fever. Some of the symptoms may point to typhoid or continued fever. The physician must be guided by the general character of the symptoms; the pain in back, debility, perspiration, and vomiting ought to make him suspicious of small-pox; and as a few hours may turn his hesitancy into certainty, it is as well that he should take the needed precautions in the way of keeping the patient separate until he sees whether an eruption is likely to appear, and if smallpox is prevalent so as to increase the likelihood of its being the disease, then vaccinating those who are in communication with the patient who have not already been re-vaccinated. Indeed where there is any good ground for suspicion, every unvaccinated person should be at once vaccinated.

The coming out of an eruption removes many doubts, but not all, as those who have much experience well know. Other eruptive fevers, such as measles or chicken-pox, may be suspected. Febrile lichen and syphilis may also cause difficulty. In measles the coryza, and the time between rigor and eruption, seventy-two hours, as against forty-eight in small-pox, must guide us. The papillae of small-pox are said to be firmer than those of measles, but, as before hinted, the *tactus eruditus* may not always be present to satisfy the judgment.

Of chicken-pox I have already spoken.

Febrile and syphilitic lichen may cause some difficulty in the early and papular stage, before a decision is arrived at. In lichen twenty-four hours of primary fever intervenes before eruption after first appearance of illness; in small-pox, when not modified, forty-eight. Small-pox appears first on face, lichen uniformly over head and trunk.

The papula, vesicle, and pustule of secondary syphilis are preceded by a febrile attack of variable duration. The pustules have a tedious course of over ten days, and "are developed, not simultaneously as in small-pox, but in successive crops." *Gregory*, p. 70.

There is a possibility of mistaking the variola haemorrhagica, or black small-pox, for purpura or scarlet fever; the passing the finger over the skin may detect the hard but otherwise undeveloped papule; or on the soft palate a few white pustules may be detected. — *Hooper's* "Vade Mecum."

Treatment. — Statistics show a lessened death rate in the treatment of this disease.

Whether this is in part owing to its being less severe than formerly it is difficult to say. It is, however, fair to infer that considerable improvement has taken place in the general management of these cases; and though allopathy may still be without its specific remedy, which would, I believe, like other specific remedies, owe its virtue to its homoeopathic action, yet in the absence of the old-fashioned violent remedies, and in the more careful attempts to meet particular symptoms by some gentle palliative, a great change for the better has taken place in the treatment of this disease. As regards homoeopathic treatment, though no one specific remedy is to be found in our Pharmacopoeia, yet as several medicines are specially indicated for various stages of the disease, I do not think it unfair to claim for homoeopathy a decided superiority. It has an equal advantage in the general management, and for particular groups of symptoms we have a law to guide us that helps us in other cases, and therefore ought to do so here.

Though some homoeopathic statistics have been given, I do not wish to press them into our service, as I know how open they are to criticism; and when I see how ready some men are to accept things as true on a very moderate amount of evidence, it is infinitely safer to do without such aid until a rigid inquiry can be instituted, that will place the matter in dispute beyond question. This was done by Mr. Marson, who showed that in the case of Sarracenia Purpurea, this medicine in no way deserved the exaggerated praise it had received. Mr. Marson tried Sarracenia in fifteen severe and likely to be fatal cases of small-pox: all died. He says: "I cannot say that I saw any effects from it whatever. It did not save' life, it did not modify in the least the eruption of small-pox, it did not influence any of the secretions; in only one case it seemed to act on the bowels, but that was doubtful." Haldane was equally unsuccessful. At one time I did urge my friend Dr. Gregory to test homoeopathy in the Small-Pox Hospital, but unfortunately he would not accede to my request.

As regards the general management, the patient should not be overheated; the atmosphere of the room should be kept as fresh as possible. Cooling drinks should be allowed freely. Cold affusion has been recommended where there is much heat of skin but I think as a rule tepid sponging is safer and more refreshing to the patient. The patient's strength must be sustained with such nourishment as can be borne and when necessary stimulants administered freely.

In considering the medicines likely to be of use, it will be well to remember the peculiarities of this disease, as shown in its various stages, and the prominent symptoms that cannot be overlooked, but must have a special remedy. These we briefly refer to, to mark the conditions under which they may be given.

1. First, then, we have rigors, weariness, severe backache, vomiting, and tenderness of epigastrium, with other symptoms of fever; and often, as a marked feature, sweating.

2. These symptoms are followed in about forty-eight hours with the breaking out of eruption, the little hard spots being readily felt by the finger. These become vesicular about the fifth day, and are followed by some abatement of the fever.

Now, there are two medicines, I think, specially indicated in this early stage —

Bryonia and *Thuja,* — Both of these medicines may help in all stages of the disease, but much depends on a careful selection throughout. And I would here caution the student to remember that the disease will be of some duration, and will go through its regular stages; therefore he must not expect to see an immediate result follow the administration of a medicine, but must rest content to know that he has chosen the best, and then adhere to it steadily, using only such other auxiliary medicines as may be called for by more special symptoms. *Thuja* is believed by some to exercise a specific influence over the disease, but in choosing between the two I have named, the symptoms present must guide our choice, and I am inclined to think that *Bryonia* will meet most of our wants at the commencement.

Bryonia. — When the patient is first attacked this medicine will be found very suitable for the symptoms that present themselves; the shivering, the backache, the headache, and the sweat are all met by it, also the vomiting and pain in stomach; therefore it will be well to begin our treatment by giving it every one or two hours. Constipation being present with adults, this medicine may be further called for if it exist, though the constipation of *Bryonia* is more of a chronic character. Diarrhoea with previous colic is an indication for this medicine, and in a later stage the bloody stools. It may help us into the eruptive stage, as pimples and vesicles and itching are among its indications, as is also bloatedness efface.

Thuja having been recommended by Boenninghausen, who used it in an epidemic in his neighbourhood, in 1849, was at once raised into a position of importance as a remedy for small-pox; indeed, so strong is the faith of some in this remedy that it has been advised to substitute it for vaccination: a very simple way of bringing a valuable medicine into disrepute.

If there are any sound reasons for not vaccinating this medicine may be used, and it is possible that like *Belladonna* as a prophylactic in scarlet fever, it may be a means of preventing the spread of the disease or making its attack milder; but to put it in place of vaccination is I think an act of unwarrantable rashness.

If it should be found on further trial that it exercises any specific action on the disease, it will of course take its place as our best remedy, but just now it is on its trial, and we can only give it its place as homoeopathically indicated.

It meets the first fever, the chill, backache and headache. The vomiting, pain in epigastrium, and sweat a little less markedly than *Bryonia*, but still sufficient for our purpose, if we think it meets the case better in any other respect, or if "its use in disease" should induce us to begin with it. And for the hoarseness that comes on about the sixth day, as well as for its action on the genital organs, it is well indicated. Therefore if we begin with *Bryonia*, and see reason for changing it, we might about the sixth day substitute *Thuja* for it, or give four doses of one, then four of the other, watching the action of each, or give *Thuja* in this way with such a medicine as *Mercurius*.

Dr. Gutteridge, who saw an epidemic of small-pox at Leicester a few years ago, recommends the use of *Thuja* as a lotion.

Aconite. — For hot burning fever, and want of sleep, this medicine should be given at night every half-hour till sleep is procured, omitting the other medicine temporarily. As a rule it is useful in all acute febrile attacks; and in all stages where feverish symptoms manifest themselves, this medicine should be remembered. In the case of children it will be well to think of *Chamomilla* or *Belladonna* for the same class of symptoms. The tendency to convulsions, which is so common among infants, as has been pointed out, would make either of these medicines, but especially the latter, of much use.

In severe illness I always advise the frequent repetition of the dose, and have no doubt that such of my colleagues as prefer giving their medicine at rather long intervals, would act wisely if they would see and judge for themselves as to the effect of more frequent repetition.

Belladonna is further indicated by the backache, headache, and delirium, and like *Aconite,* it may be of use at various stages: the swelling of throat and glands; and the erysipelatous as well as the gangrenous character assumed at times, may call for its administration.

Croton Tiglium. — The very powerful action of this medicine on the skin; the swelling, erysipelatous inflammation, vesicular and pustular eruption, and great irritation; the exhaustion, sore throat, ophthalmia and purging, make it desirable that this medicine should receive more attention than it has done. I consider the excessive irritation a strong reason for its use, judging from what we know of its curative action in other diseases characterized by this.

Ipecacuanha may be required for a few doses if there is excessive vomiting.

Antimonium Crudum is a medicine that should be thought of as well as *Ipecacuanha* for excessive vomiting. It also may be of use for the feverish symptoms and drowsiness, and to some extent for the eruption; on this account it is well to give a medicine such as this in preference to *Ipecacuanha* if the greater number of symptoms are to some extent covered by it.

3. Following the progress of the disease we find, about the sixth day hoarseness, swelling of salivary glands, and the following day the contents of the vesicle are becoming purulent. These and the other symptoms, lead us to consider the desirability of a change in our medicines.

As I have mentioned, *Thuja* may be of use at this stage, but we must now briefly direct attention to some other medicines that may also be called for.

Kali Bichromicum is characterised by an itching eruption like small-pox, of a pustular character; soreness of throat and dryness, cough and irritation of larynx, ulcerations of mouth and coated tongue.

The eye symptoms of this medicine would point to its use at a later period, when this organ is in danger.

Antimonium Tartaricum. — The fever, profuse sweat, pustular eruption, gangrenous ulcers, prostration, and failure of pulse, soreness of mouth, difficulty of swallowing, diarrhoea and bloody stools, make this medicine of use especially about the eighth or ninth and following days and at a later stage; it is one of the medicines that may be given in any form of small-pox, though not without auxiliary help from some other medicines in black small-pox.

Kali Nitricum. — Having a vesicular and pustular eruption and profuse sweat, comatose sleep and wandering, and bloody stools, though not at all equal generally to some of the other medicines given, is worth consideration as one to consult with the aid of our repertory.

Stramonium. — Dr. Hartmann, in his excellent work on the treatment of acute and chronic disease, a book now a little out of date, speaks very highly of *Stramonium* as having been frequently useful in accelerating the appearance and shortening the course of the pocks. Among its symptoms are convulsions, drowsiness, blisters on skin, itching eruption, swollen face, turgid with blood; swelling of face, eyes, and tongue; ptyalism; hoarseness; delirium; tremulous, weak, unequal, sometimes intermittent pulse; small quick pulse; extinct pulse; at another time the pulse may be strong and full; sweat; difficult deglutition; spasm of oesophagus; suppression of urine. From these symptoms our recommendation of Stramonium may go beyond Hartmann's, as it may be serviceable as a direct antagonist, or as a useful ally to some other medicine at most stages of the disease, especially for violent delirium.

Veratrum Viride is recommended by different writers as equal if not superior to Aconite in the first stage of small-pox. Dr. Holcombe, of New Orleans, bears testimony to its value, but the evidence in its favour is as yet far from complete; nevertheless it is a medicine of such promise, that it should not be overlooked where the opportunity occurs for testing its efficacy.

It would be interesting to see if the new medicine *Jaborandi* which acts so powerfully in producing salivation and sweating, would be of use as an auxiliary remedy, where it was thought desirable to control these symptoms.

Mercurius Corrosivus should be given in preference to *Mercurius,* if better indicated, as it will be where some of the aggravated symptoms spoken of are present; thus bloody urine, bloody dysenteric stools and ptyalism may make it a valuable remedy either alone or in conjunction with such a medicine as *Rhus.*

If we knew more of the action of this medicine we would probably find that it might often be substituted with advantage for the milder preparation of Mercury.

Opium. — In cases of suppressed eruption, in profound coma, in all stages where an effort is needed to rouse the patient, a few rapidly repeated doses of this medicine may be of use. It might be given every fifteen or thirty minutes for six or eight doses.

It meets many symptoms, though not the whole group; but enough to make it of increased value for the purpose for which it is given.

4. As the pustules are coming to their height, which they reach about ninth day, this period is usually preceded about eighth day by the setting in of secondary fever, which is often very severe.

In confluent small-pox the maturation is earlier, and the secondary fever much more severe, assuming the character of typhus. Petechiae, haemorrhages,, sloughing, dysentery, coma and convulsions, have all to be provided for in the treatment as they arise.

Arsenicum at this stage, as well as in black small-pox, is a medicine we must rely on. The hoarseness, angina and diarrhoea are met by it; the character of the pustules as seen in confluent small-pox, the blotches filled with blood and pus; the bloated puffed face, as well as its opposite, the sunken countenance and pale death-coloured face, the bluish lips, the brownish blackish trembling tongue, the aphthae, fetid smell from mouth, loss of consciousness and loss of speech, stupefaction or muddled state, pains in stomach, vomiting, the dark putrid stools, or dysenteric diarrhoea, the inflammation of eyes, the chill or heat of fever, the languor, uneasiness at heart, delirium, restlessness, grasping at flocks, the sweat and haematuria, the spots resembling petechiae, bluish spots and tendency to putridity all combine to show the value of the remedy.

If we have commenced our treatment with *Bryonia* or *Thuja,* and followed with *Mercurius, Arsenicum* may come in during the progress of the secondary fever. Of course *Belladonna, Stramonium* or *Hepar* may have been needed; but unless specially called for. the fewer medicines given the better.

Bryonia. — The dryness of mouth, coated tongue, indistinct speech, delirium, petechiae, red, brown and scanty urine are indications of this medicine, which has been named before for the early stage.

Rhus meets the fever, erysipelatous redness, petechiae, black pustules, snoring, grasping at flocks, confusion of head, dry, red, parched tongue, diarrhoea, drops of bloody urine, and dark urine. This is a very valuable remedy, and in a severe case might be given after *Arsenicum,* four doses of one, then four of the other.

Where medicines are thus given the nurse should be cautioned to give a few more doses of the medicine that seems to give most relief, or to adhere to it entirely. *Baptisia,* though much thought of by some, has not yet acquired that general approbation that the other medicines have. Fevers with gastric disturbance, heavily coated tongue, and muttering have been benefited by it. Hale, in his New Remedies, tells us what is known about it. It is not likely that in the fever accompanying small-pox it will ever equal *Bryonia* or *Rhus,* still some symptoms may call for its use.

Mercurius. — Though some of the other medicines I have named may prove of great value, yet perhaps, as a general rule, there is no medicine better suited for the suppurative stage than Mercurius. If we have begun our treatment with *Bryonia,* then followed it with *Thuja,* as suppuration goes on, we may find *Mercurius* next called for. The eruption, swelling of cheeks, nose, swelling of genital organs, salivation, soreness of throat, swelling of glands, point to this medicine. Boehr says, "Just as well as *Mercurius* will prevent suppuration in cases of abscesses and small boils — and we know from abundant experience that it will do this — we may likewise expect a favourable effect from it upon the suppurative process in small-pox, so much more as the symptomatic similarity between the pathogenesis of mercury and smallpox in this special point extends to the minutest particular." As before suggested, we may find it desirable to give four doses of *Thuja,* then four of *Mercurius;* or if we have delirium. *Belladonna* or *Stramonium* at night, or it may be needful to aid or follow its action with some other medicine. *Hepar Sulphuris* is likely to be of use in the suppurative stage. The eruption, erysipelatous swelling of eyes, soreness of throat, and bloody urine are indications for it. It would be well that a little more attention were paid to this valuable remedy in confluent small-pox.

Other medicines, such as *Acidum Nitricum* or *Apis* may be of use when there is sudden accession of erysipelatous swelling, or *Hyoscyamus* for teasing cough; this last may be given rapidly for a few doses, to relieve this special symptom. If it should fail in a higher dilution I would recommend a trial of 1x for a few doses before its use is abandoned. Though homoeopathic to the cough, it would still be as a homoeopathic palliative rather than as covering the symptoms generally that we would use it.

Cicuta Virosa is homoeopathic to head symptoms, such as sleeplessness, stupefaction, delirium, and convulsions, also retention of urine; and as it has suppurative eruptions on face, and itching of the whole body, with confluent pimples on hands and face, it may be classed amongst our auxiliaries.

Gomodadia Dentata is another medicine of this class. It is chiefly indicated by the itching, and erysipelatous swelling of face, and inflammation of skin, followed by purulent and offensive discharge from the deep ulcers that form.

Lachesis. — Though it does not take in some of the prominent symptoms that we should wish, meets some others so well, and is of such known value in allied diseases, that we may look upon it as a very valuable auxiliary medicine. Drowsiness and head symptoms, tossing about and moaning, starting when falling asleep, convulsions, erysipelas and gangrenous blisters, typhus fever, or symptoms corresponding, small unequal pulse, delirium, unintelligible muttering, black tongue, swelling of face, as well as. isolated symptoms that are found in its proving, should lead us to the use of this medicine. It may act well following some other medicine, but always remember if a remedy is acting satisfactorily not to disturb it, and where possible the more we can keep to a single one the better for our patient, though in a disease of such violence, and that witnesses so many stages, there is more reason for a change than in other diseases where a more continuous course is the rule.

Acidum Muriaticum claims attention as a medicine that is worth further consideration. The black pock, putrid state of ulcers, oppressed state of head, disturbed sleep, typhoid character of fever, epistaxis, and salivation, entitle it to a place as an auxiliary. Though it does not take in a full group of symptoms, still medicines of this class must not be discarded, as some of their characteristics may from time to time raise them to the rank of valuable curative agents.

Secah Cornutum. — This medicine is of greater value, as its known action gives it a position as an agent for combating the state of degradation of the vital powers met with in severe cases, either in the variety of disease known as black small-pox, or in the advanced stage of confluent, where petechiae, purpura, and gangrene require some powerful antagonist. The convulsions, the black suppurating blisters, bloody blisters assuming a gangrenous character, the ecchymosis, loss of consciousness, somnolence, stupor and restlessness, diarrhoea, haemorrhage, feeble intermittent pulse, and laboured respiration, are strong indications for its use. It may be given alone, every 20 to 60 minutes, or six or eight rapidly repeated doses, after an equal number of doses of *Arsenicum* or some other remedy.

Sulphuric Acid. — The small quick pulse, subsultus tendinum, ecchymosed spots, and the recommendation it has received in purpura, make me class this medicine with one or two others that I have named for consideration in some urgent cases.

Terebinth. — The value of this medicine in haemorrhage gives it a place in the treatment of small-pox, whether in the form of variola haemorrhagica or of confluent small-pox where haemorrhage supervenes. The eruption is slightly characteristic of this form of the disease, as it is apt to appear and disappear, and though erythematous is also papular and inclined to be vesicular. I do not, however, attach the weight to this that I do to the haemorrhagic tendency.

As regards the strength, though it may act well in a high dilution, there is a greater amount of evidence of its action in the more tangible form. It may be given in the strength of the 1x or 2x dilution in water.

Arnica. — The jactitation and twitching of muscles, convulsions, sinking of strength, ecchymosed spots, hematemesis, haematuria, epistaxis and head symptoms, are our reasons for selecting this as one of our medicines in variola haemorrhagica, and as an auxiliary in the confluent variety. It may be given in the same way as *Secale*.

Camphor. — Must be looked on as an auxiliary remedy, as except for collapse, erysipelas, torpor, with loss of consciousness and convulsions, it is not otherwise directly indicated, but for the above as well as isolated symptoms it is a valuable medicine. Camphor is one of the few medicines that is almost universally given in larger doses than our other medicines. Some caution is needed, as in the directions given by the chemists the danger of an over-dose is not pointed out and as some persons are very fond of domestic treatment, they at times buy their experience rather dearly. It seems to act well, feiven in small but rapidly repeated doses; a dose every ten minutes for six or eight doses. It may be needed longer, but a medical man will use his judgment as to that, and also as to the quantity given. [2]

For the purposes required, camphor will be best given in the 1 in 6 strength, or this strength slightly diluted; but camphor has been given in the diluted form, and chosen strictly in accordance with its recorded action; there is no reason why it should not act as well as other diluted medicines. Thus, my friend Dr. Wilson, selecting it in an urgent case where a child was suffering from cerebral disturbance, rotatory movement of the arm being the key symptom, gave it successfully in a diluted form.

The existence of this symptom in the stupor of fever or some other disease, would point to its use.

There are two other remedies that have been recommended, that I ought to name, though I confess to a natural repugnance to their use, and should not myself feel disposed to employ them unless they prove to be of such value that it would be wrong to lay them aside. At present I think there is no such conclusive evidence in their favour.

Vaccinium is said by Raue, in his Pathology and Therapeutics, to have shortened and ameliorated the disease. *Sulphur* was given after it. Used in the form of vaccination, especially in the early stage of incubation, there is a hope that the disease may be modified by it; later, it is difficult to say how far it produces an effect, the two eruptions will go on concurrently at times. Dr. Harmar Smith, who has had some experience of vaccination as a remedy, thinks well of it.

Variolinum, Raue says, has been equally successful, and that it has the testimony of ten physicians in its favour, who used it in different epidemics. He does not specify the strength in which it was used. Dr. Dudgeon treated a boy aged six with this, he took some lymph from the pocks, and with diluted alcohol formed the 1st dilution, of which he was ordered to have two drops every hour while awake. The face was painted with *Collodion,* to which the boy objected strongly. The case did well. (*British Journal of Homoeopathy,* vol. 10, page 262.)

5. Treatment of Variola Haemorrhagica. — This terrible form of the disease (the only one requiring any separate notice), as already stated shows its malignity at a comparatively early stage, and life is terminated with less distress than in the confluent form when it is fatal. To meet this our measures must be prompt and decided. I need not go over the indications, it is sufficient to name the medicines, and refer back to what has been already said as to their action.

The non-appearance of the eruption should lead us to select *Ipecacuanha* or *Opium,* the former where vomiting or any threat of haemorrhage was present, the latter where there was drowsiness or a generally depressed state. A few rapidly repeated doses of the medicine selected should be given. This might be followed by a few doses of *Sulphur,* after which the former medicine might be given, the object being to combat the disease and secure the throwing out of the eruption, the *Sulphur* being calculated to help in this. After some hours of this treatment, the appearance of the more active symptoms should then lead us to choose from among the medicines named.

We probably will find *Arsenicum* of use as one of the early ones, and as already advised, following every six or eight doses of it with *Arnica* or *Secale. Phosphorus* is indicated by the haemorrhagic tendency, the cough, sudden and excessive weakness.

If haemorrhage becomes more prominent than the ecchymosis and petechiae, then Terebinth may be brought into use. *Hamamelis* is a medicine that ought to be of some benefit, but I am not aware that it has been used.

If symptoms of collapse are present, or if the eruption begins to dry up, *Camphor* should be at once thought of, and be given rapidly, if selected in preference to *Arsenicum* or *Phosphorous.*

There is yet one class of medicines that ought to be studied in connection with each case. I allude to the snake poisons. *Lachesis* I have spoken of, but there are others well worth our consideration as auxiliary medicines, especially two, the *Vipera Redi* and *Vipera Torva,* the Italian and German vipers, I should like to see more used in severe forms of the disease, either confluent or haemorrhagic.

I would venture to give a caution to my brethren, to be slow in coming to the conclusion that medicines are of value because they appear to act well in one or two severe cases or in many mild cases. Let them note what they observe, and if they find the same thing: repeated in severe cases, they may gradually accept the evidence. With mild cases great care and discrimination are needed, or we may have remedies puffed in a very dangerous manner.

This has been done; and if, unfortunately, done by a man of some standing, the mischief is not easily undone.

General Management. — There is much to be done besides administering medicines.

During the progress of the disease the patient's strength must be sustained by food and stimulants, and cooling drinks should be allowed freely. The air of his apartment and of the house generally, must be kept fresh and pure. Condy's fluid should be used in the sick-room, and cloths wetted with solution of carbolic acid hung up in the passages, and the clothes of those going about the sick person played on with carbolic spray. We must, however, take care that, in our zeal to prevent the spread of the disease, the patient's medicine be kept pure and free from being destroyed by the carbolic. I do not think patients always get sufficiently thought of by nurses and others in this matter. Our medicines are very delicate, and may readily be injured; therefore we cannot be too careful to keep them in well-corked bottles, or in tumblers *covered with white unscented paper,* and a book laid outside this to keep it well covered; the spoons also must be kept separate and clean, and should not be wiped with the first towel that comes to hand, and which may have carbolic or Condy on it. These matters seem trifling; they are vital to the patient.

All unnecessary furniture and carpets should be removed from the sick-room. Quiet should be secured as much as possible. To relieve the itching, which is very irritating, nothing is pleasanter than the application of the *Linimentum Calcis,* applied by means of a brush or feather. Simple ointment, or this with the oxide of bismuth added, is used by some, while others apply starch powder.

To Prevent Pitting. — Various measures have been employed; the fact that no one method is much preferred rather tells against all. Mercurial ointment has been used with some apparent benefit, but as this is absorbed, and may produce its constitutional action, it is inadmissible in homoeopathic practice. Simple olive oil has been used, also collodion; this also is applied mixed with equal parts of castor oil. Nitrate of silver and iodine have been applied to the pustules with some apparent benefit, but this is attended with much pain, so that it should only be used for a few special pustules. *Calamine* (carbonate of zinc), mixed with olive oil to form a crust has been recommended by Bennett. Another method, and the least objectionable, is to puncture the vesicles with a fine needle and take up their contents on cotton; this is a tedious process, but if it will lessen the disfigurement that is a small matter. This and the use of the *Linimentum Calcis* are probably the best measures within our reach.

After Treatment. — The affection of the eye coming on in the progress of the disease has been spoken of, and *Kali Bichromicum* suggested as a possible remedy. *Mercurius Corrosivus, Hepar, Phosphorous, Arsenicum,* and *Sulphur,* are medicines also to be looked to. Ophthalmia may follow small-pox. This must be treated by the above medicines, to which list we may add with advantage *Euphrasia* and *Euphorbium*.

Inflammation of the testicles may be treated with *Clematis, Pulsatilla, Nitric Add, Mercurius, Belladonna, Aurum*, if they remain indurated. For simple swelling I have used a *Conium* lotion with great advantage.

If the ovary is affected *Conium, Apis, Zachesis, Belladonna, Bryonia or Gantharis, Platina* may be of use in some cases.

As far as possible all the symptoms must be taken into account in making our selection.

Abscesses will need *Belladonna, Hepar, Phosphorus,* and *Silicea*, and the application of poultices, and where the bones are affected, such medicines as *Aurum, Calcarea, Asafoetida, Mercurius* and *Stillingia*.

General debility will need good nourishing diet, and such medicines as *China, Muriate of Quinine,* [3] *Ferrum, Phosphoric Acid,* besides such other special medicines as are indicated.

Besides the above remedies, change of air should be resorted to as one of the best restorers we have.

Warm baths should be used for the comfort of the patient, and freely for the protection of others: too much care cannot be taken before the patient is allowed to mix with other people.

The patient's room after the illness is over will need thorough purification, washing, removal of the old paper, re-papering, and whitewashing.

[1] Note – Since these lectures were delivered, we have been visited with another epidemic of small-pox, in which the deaths have been over one hundred in a week.
[2] Dr. Quin's preparation of Camphor 1 in 6 is sufficiently strong for all purposes, and was the one in common use in this country till what is known as Rubini's Camphor came to be asked for more frequently. Hahnemann originally used 1 in 12. This confusion is to be regretted. Were it not for the difficulty of displacing the preparations in use a strength of 1 in 10 would have been the best officinal preparation, as it is, the 1 in 6 strength is the one that the chemist is directed in the 1876 edition of the British Homoeopathic Pharmacopoeia to give, unless some other strength is ordered.
[3] This is more soluble than the sulphate, and I believe has a very similar action, both remedies, however, want farther proving and clinical observation.

Inoculation

It is a curious circumstance that the Turk, who carried destruction and desolation in his path, should have been the instrument of making the most valuable medical discovery the world ever witnessed; but so it is, for about A.D. 1700 inoculation was first practised in Constantinople. In 1717 Lady Mary Wortley Montague had her son inoculated in that city, and in 1721 her daughter was operated on in this country. She was thus instrumental in introducing to our nation the means that had been placed within our reach for arresting the progress of the dread disease of small-pox. Like many other benefactors she had to endure no small amount of annoyance; let us hope that she found a reward in the consciousness of the good she had done.

The same year that saw it introduced here also witnessed its introduction into America.

Its advantages gradually became apparent. The Princess of Wales had her two daughters inoculated, and this example led the way for its more general adoption, and in 1746 the Small-pox Hospital was established to extend the benefit. In 1754, thirty-three years after its introduction, and about fifty-four after its discovery, it received the sanction of the College of Physicians. [1]

It is difficult to explain why small-pox as introduced into the system by inoculation should be less severe than if taken by coming within infectious distance of some person suffering from it. Perhaps a smaller dose of the poison is administered than where a large exposed surface of the body, as the skin, or the lungs is the medium through which it enters; but as the fact is so, those that wished to avail themselves of it had the benefit, till a still milder process took its place. Inoculation had become so trusted that it required an Act of Parliament to displace it, and no doubt many were inoculated even after the passing of the Act.

Inoculated Small-pox. — When inoculation was practised, the lymph used to be taken from the vesicle on the fifth day. and was inserted by a single puncture into the arm. The lymph taken thus early was preferred to the purulent matter taken on the eighth day from a pustule, which would have been equally efficacious.

When successfully performed, the second day after the operation a lens will show a slight orange stain; the fourth or fifth day the part is hard, slightly inflamed, itches, and a vesicle is formed on it. About the sixth day some pain and stiffness is noticed in the axilla, on the seventh day the vesicle is more formed, with an areola. From the seventh to the ninth day fever appears, lasting three or four days, and is followed by a general eruption, which is called the secondary eruption. Like regular small-pox, the eruption comes out in three crops, appearing successively on the face, the trunk, and the legs. The original pustule has scabbed when the secondary or general eruption is about to maturate. In some cases three or four vesicles appear and shrivel up without maturating. The number of spots may vary from ten to two hundred. The induced disease is usually mild, but in some cases severe, the eruption confluent, and death has ensued. The possibility of this is one of the objections to the operation, but the danger of spreading the disease has induced the Legislature to forbid the performance of the operation.

I am chiefly indebted to Aitken's "Practice of Medicine" for the above particulars. Those who have witnessed inoculation have placed on record the phenomena attending it, otherwise we might remain in ignorance of what took place, as a month's imprisonment might be the fate of any one wishing to witness the facts for himself.

That an inoculated individual may become a centre of infection, and that the disease has been spread in that way is very possible, but the opponents of the system no doubt made the most of their case, and in blaming inoculation may have overlooked the fact that epidemic influences would also spread the disease. We have now no inoculation, but we have plenty of small-pox, showing that some other agency must be at work causing it to spread.

The possibility of inoculation causing death, and vaccination being so much milder, and being believed to afford efficient protection, was an argument, as against the one, and in favour of the other, that could not be withstood; the object being to obtain the greatest possible security at the smallest possible risk. Small-pox contracted naturally or by inoculation no doubt affords the best security we can have against subsequent exposure to that disease; but if it can be shown that vaccination efficiently performed, and repeated at an interval of some years, will secure the same immunity, the milder and safer operation must still be preferred.

That vaccination, *as it is,* fails to give the confidence that it ought cannot be denied; that it may be made vastly more efficient there is abundant evidence to prove.

When inoculation was discontinued it was open to very grave charges. Precautions were not taken as they ought to have been, to prevent the spreading of small-pox, by means of it; and every one that liked operated. Had the Legislature restricted the performance of the operation to properly qualified persons, who should have superintended the isolation of the patient, much of the objection to it would have been got rid of. Permission might also have been given for the inoculation, under special circumstances, of persons who had previously been vaccinated. The late Dr. G. Gregory (I know from what he told me), rather favoured this idea, if the law would allow of its being done. Of course a certain number of years should have been allowed to elapse after vaccination, to give a chance of the inoculation taking, which it might be expected to do in a very mild form. But the most perfect isolation would still be required.

It is not likely that inoculation will again be resorted to in this country, though unless vaccination is greatly improved, and can be made to give greater security than at present, a demand might arise for an alteration of the law, so as to permit its being done under proper restrictions as those mentioned above.

Circumstances might arise that would justify its use in our colonies. Thus, if an outbreak of small-pox took place unexpectedly, where a supply of vaccine could not be obtained without delay, inoculation would afford the readiest means of protecting those exposed to danger, and would most probably be at once resorted to. Lymph could be procured by the means named earlier, but as there would be delay in thus creating a supply, it would not be wise to wait in localities where the disease was spreading rapidly. If such an unhappy state of things did arise, isolation of those operated on should be strictly enforced.

An emergency of this kind has arisen; but, unhappily, the knowledge that a remedy was within reach was not thought of, and that a fresh supply of vaccine might be obtained in a comparatively short space of time was unknown.

Mr. Henry M. Stanley, the discoverer of Livingstone, in a letter dated Ujiji, August 13th, 1876, which has appeared in the *Daily Telegraph* while these pages are going through the press (March 29th, 1877), describing the position in which he found himself owing to a fatal outbreak of smallpox, which had caused great mortality, thus writes: "You may imagine, then, the feeling which prevails in all minds at the present time in Ujiji — it is that of dismay and terror; and as the inhabitants look forward to two months more of the fatal experience they are now undergoing, those who are able to quit the horrible spot should pack up at once."

Speaking of his own difficulty he says: "The only thing it appears to me that has saved the Expedition from total wreck is vaccination. But I find, when too late, that many of the people lost the benefit of this precaution from sheer laziness — when summoned they would not appear. My vaccine matter is all dried away now, and not a particle of it can be scraped up to be of use." Inoculation here would have been the remedy, and if, at the same time, a few cows were inoculated, a supply of vaccine might have been obtained so soon as the operation on one or more of the animals had been attended with success. After this the vaccination might have been used from arm to arm. Had Mr. Stanley been acquainted with the means so ready to his hand, and been instrumental in introducing such a blessing in that part of Africa, he would have paved the way for trade, civilization, and, above all, for missionary work.

Probably, in the event of such an epidemic appearing in any colony many would be found who, having been vaccinated in the mother country, would be safe, and from amongst these attendants on the sick might be found. Doubtless, however, some of them would seek further protection in inoculation. In our hospitals here revaccination is found to afford perfect security to the nurses. The great drawback to it is the need for its repetition, whereas inoculation, unless in rare cases, will secure immunity during life.

[1] This learned body has not been noted for its rapidity in welcoming great discoveries. It still frowns upon Homoeopathy, and shows but little inclination to test its claims, though individual members of the old school do not scruple to use the information they get from us. In doing this they pay us a compliment, and as the public benefit by it, we may be content to bide our time, feeling sure that the facts that are in this way established will ultimately pave the way for our vindication. At the same time I do not join in a wholesale condemnation of our opponents. There is much that is startling connected with our system. The doctrine of like curing like is not so very difficult to accept, but the small dose question as it has been presented, is one that I think should make us more considerate of those that oppose it. The small dose has come to be connected with homoeopathy, but is really not part of it, though many attach much importance to it. Of late years it has not

been put forward as prominently as formerly, and if medical men can be got to look into the general bearings of the case, I think that they will come in time to understand it better; and if led to test it in practice will see it is not so untenable as they suppose.

Vaccination

Popular ideas are not always popular errors, and it may happen that a valuable truth lies very near the surface in some of these commonly believed opinions. Thus it was well known in Prussia and Gloucestershire that persons who milked cows were apt to contract a disease on their hands from sores on the teats and udder of the animal, which it was found prevented them taking small-pox when exposed to infection, while their less fortunate neighbours suffered.

Dr, Edward Jenner, whose name will go down to posterity as the discoverer of vaccination, had heard the country talk on the subject, and when in London had spoken to the famous John Hunter about it, but failed to excite that interest in his mind that might have been expected; nothing daunted, on his return to the country he investigated for himself, and made the discovery that followed, his own. It occurred to him to utilize what his experience had taught him, and endeavour, by propagating the disease from one person to another, to give those so operated on the protection that the milkers had acquired without seeking for it.

Our colleague, Mr. Hands, of Hammersmith, who had the honour to be Jenner's pupil some fifty years ago, tells us, among other interesting facts, that it was during the holding of Berkeley fair, May 14, 1796, that James Phipps, a boy aged eight, became the first subject of vaccination from matter taken from the hands of a milkmaid; and that subsequently when tested by inoculation no result was produced, proving the success of vaccination. [1]

Mr. Hands tells us, in his pamphlet on "Vaccination and its Opponents," how particular Jenner was " relative to the condition of the child to be operated on, and still more circumspect as to the infant from whom he procured the virus, which was afterwards used for the purpose of vaccination."

I consider it unnecessary to go into any details to prove the value of vaccination, and shall accept the almost universal belief as to its utility as fully warranted. We have only to compare small-pox as it was, with small-pox as it is, notwithstanding its appearance as an epidemic, to see what a blessing vaccination has been to mankind. If those gentlemen who do so much injury by trying to run it down would use their efforts to get it improved, they would find more adherents, and might do some good.

Origin of the Disease in the Cow. — It is of importance that we should if possible obtain correct views as to how the disease began in the cow, for the double purpose of knowing how to keep up our supply of vaccine, and also to see whether what holds good in one case will do the same in another, and whether the progress of other diseases may be arrested by somewhat similar means. [2]

In the absence of direct and positive evidence as to how the disease originated in the cows in Gloucestershire and elsewhere, which of course we can never have, we must be content to draw our conclusions from the facts that come under our notice as to the source from whence the disease may have sprung. Judging in this way, I believe cowpox to be a disease that has been contracted by the animal from some individual suffering from smallpox, but that in consequence of its being taken by an animal it has undergone such a change or modification as to become to all intents and purposes a new disease, and that this new or modified disease, when conveyed back again from the cow to the human subject, produces in the individual *not what is known as modified* small-pox, but genuine cowpox, and that this will protect the individual as efficiently as any lymph can from an attack of small-pox.

When the cow is inoculated with small-pox matter, the virus from the vesicles produced in her, will in the human subject give rise to a disease in every way identical with what we know as cow-pox, and from it we can go on propagating the disease from arm to arm.

Mr. Ceeley, of Aylesbury, one of our highest authorities on the subject of cow-pox, successfully inoculated the cow many years ago. [3] "This same gentleman has also often re-communicated the vaccine disease from man back to the cow (retro vaccination, as it has been called), and he has observed that good human lymph, when re-transmitted in this manner, loses some portion of its activity. The phenomena appear later, smaller vesicles are produced, but ultimately *after successive re-inoculation* in man it regains its activity. Human small-pox has also been transmitted through the horse to the cow, and so to the child in the form of cow-pox. (Fletcher.)

Dr. Jenner believed that the two diseases were identical Mr. Ceeley gives some striking evidence of the spreading of the disease amongst cattle otherwise than by inoculation; and Aitken, who in his "Practice of Medicine" collects a number of facts to show how small-pox has been conveyed to cattle, thus sums up: — "These and similar facts seem to lead to the conclusion that small-pox and cow-pox are not dissimilar diseases, but are identical in their nature." (*Aitken,* 5th edition, vol. i., p. 271.)

It is a well-known fact that a disease affects the heels of horses known as grease, that will produce a disease identical with cow-pox; there is, however, a spurious disease, that will not give the same results.

"Fontan relates that some mares being affected with a pustular eruption called grease (eaux aux jambes), the matter from the pustules was inoculated on the teat of a cow, where it produced several fine pustules; from these several infants were vaccinated with the result of producing perfectly characterized vaccine vesicles. Thirty infants have been vaccinated from this source at Toulouse, and in all the re. suit has been most satisfactory."— (New Sydenham Society Year Book, 1860, p. 146).

Mr. Badcock, who resided near Brighton, was in the habit of supplying medical men with lymph from cows that had been inoculated: large numbers of persons were vaccinated from this source. It was said that he had to discontinue the practice, as it was illegal; this is not the case, he was obliged to discontinue it owing to an Act of Parliament forbidding the removal of cattle during the cattle plague. The obstacles that then stood in his way of procuring a supply were too difficult to be overcome. Others have obtained lymph in this way, and during the last outbreak of small-pox I was enabled to get some of this lymph three removes from the cow: from which I had some very beautiful arms.

"In 1836 Dr. Basil Thiele, of Kasan, in South Bussia, successfully inoculated some cows on the udder with the *virus* of human small-pox. Vesicles were produced, bearing all the characters of the true vaccine vesicle in those animals. The lymph so produced from the *variolation* of the cow continued to retain the specific properties of the vaccine variola throughout seventy-five successive transmissions in the human subject. In 1838 M. Thiele repeated this interesting experiment with a similar success." — *Aitken,* vol I., 270, 5th edit.

The opinion that small-pox may be spread in this way, as has been asserted, is, I believe, entirely an error; out of the numbers that have been thus vaccinated in this country we have not heard of small-pox arising from it; on the contrary, I believe the lymph so obtained to be very superior to the old stock that has been so long in use.

Writing lately to a friend who supplied me with the lymph I allude to as having used in the last epidemic, for more information, I received a letter from which I take the following: — "The cow referred to was not operated on by me, but by the farm bailiff. I did not see the cow at all. He took lymph eighth day, and vaccinated a calf inside of thigh, and not on udder; the vaccine therefrom derived was extremely active. I had some bad arms, but nothing which produced permanent bad effects."

Were the lymph taken from the first animal, to be passed through one or two, before returning it again to the human subject the arms might be less severe, and if then, strong and thoroughly healthy babies were selected for being operated on the supply from them could probably be as mild as the ordinary lymph, while we would expect it to be far more efficacious.

Besides the argument in favour of the disease in the cow originating always from small-pox, which these facts I think demonstrate, we may draw another from the fact that, almost as a universal rule, small-pox is taken but once during life, therefore it is not a very strained argument to say that a disease such as cow-pox, that will in some cases protect the individual subjected to it during life, and almost in every case for some years, must have a very intimate connection or identity with it.

Measles protects against measles, scarlet fever against scarlet fever, whooping-cough against whooping-cough, and cow-pox or small-pox against smallpox, but they do not protect against one another.

Mr. Simonds, of the Veterinary College, in 1862, during an epidemic of small-pox amongst sheep in the counties of Hants, Wilts, and Dorset, successfully inoculated the animals with the matter the disease supplied.

There is a case taken from the *Boston Medical and Surgical Journal* referred to in the New Sydenham Society Tear Book for 1860, that would be evidence in favour of the disease being spread by inoculating the cow, if it were reliable, but it proves a great deal too much in the proportion of the deaths, and is wanting in details; it is as follows:—

"Martin inoculated some variolous matter, taken from a pock upon the body of a man who died of variola, into a cow's udder, and subsequently vaccinated about fifty persons with matter derived from the cow; most of those so inoculated were attacked with variola, and three died."

The author of this rash experiment may, very possibly, have transmitted something more than small-pox virus to the cow, or he may have transmitted the disease otherwise than by his lancet, as this one experiment is entirely at variance with our experience here.

Cow-pox in the Cow. — During the prevalence of small-pox, cow-pox may show itself, and horses and sheep are liable to be attacked as well as cows, and sometimes with much violence.

It would be hardly prudent to seek a supply of vaccine during the prevalence of a malignant variolous epidemic amongst cows; though at its decline, the time at which it is thought Jenner, probably unwittingly, obtained his supply, we might expect w get lymph of a sufficiently mild character to give security, without producing severe arms or other mischief.

We can readily understand how the disease may spread from cow to cow through the medium of the milkers, and it requires no great effort of the imagination to understand how some one sickening for small-pox may originate it.

The description we have of the disease is, —

There is heat and tenderness of teats and udder for three or four days, then irregular pimply hardness. The pimples get red when about the size of a pea, and increase to the size of a horse bean. The vesicle becomes a globular oval, and ultimately an acuminated pustule. "A central depression with a marginal induration is the form ultimately assumed, and when punctured towards the centre the vesicles yield a more or less viscid amber coloured fluid. Brownish or black scabs then form, and when these are rubbed off a slight central slough is seen at times. These appearances may be seen in various stages, showing a succession of crops." The period of incubation after casual communication of the disease seems to be from six to nine days, although it is said pimples may be felt under the cuticle the fifth day. The mature vesicle measures eight to ten lines in largest diameter, the centre and edges of the intumescent margin being of a deep blue or slate colour, and the surrounding areola of a pale rose, and seldom more than four or five lines in depth, the integuments under it being deeply indurated; the abundance of lymph makes it assume a conical shape. The succeeding crust or scab comes off from the twentieth to the twenty-fourth day. A vesicular eruption coming out in crops is apt to appear about the ninth day.

"The best lymph is obtained from the perfect vesicles near their centre, before they begin to point. Later than this it is less reliable. Vesicles on which the central crust have begun to form are most productive; small superficial vesicles are often productive; the great object is to get the pure lymph, free from blood. The disease may be propagated by 'amorphous masses of concrete/ slightly amber coloured lymph, found close to broken vesicles, also from the central conical dark brown crusts, and from desiccated vesicles.

"These dry materials may be reduced to powder, and then moistened with Glycerine." — *Aitken.*

When the human subject is vaccinated from the cow it is more apt to take sharply than when done from arm to arm.

Operation of Vaccination. — The manner in which vaccination is performed is of some importance, every one has his favourite method. My own practice is to make some very fine cuts or scratches on the arm with a lancet barely enough to show a slight red stain, and then to rub in the lymph off the instrument, flat glass, or points, trying to get five vesicles. If much blood flows the operation is pretty sure to fail Should it not succeed, it should be repeated, and if necessary more than once.

Instruments have been made to facilitate the operation, one like a little rake for scratching, another for cutting to a limited depth, but those who are expert with the lancet will not care to use them as they cannot be so easily cleaned — a matter of no small importance.

I have never met a case of insusceptibility in a child, and am sceptical as to the existence of such; I believe it is only previous vaccination or having had small-pox that will prevent success; even those that have been vaccinated before will, after a few years, be so far susceptible as to show some result in the shape of abortive vesicles, even if they do not take fully. Therefore I always try two or three times in case of re-vaccination, if I do not succeed at the first attempt; and as I do not. know whether the fault is mine, or that the lymph is inoperative, the second time I use lymph fresh from the arm if I can get it. I recently vaccinated a lady successfully from her own baby, who had been operated on in all eleven times, including infancy, unsuccessfully. She told me afterwards that she would not have submitted to have it done, only that she expected me to fail, and on that supposition was done with other members of her family.

Human Vaccination. — When an infant is successfully vaccinated, there is a little elevation the *second* day, but as this may also result from the scratching or puncturing it is impossible to tell with the naked eye at this early stage if it has succeeded; a lens may show a slightly vesicular appearance; about the *fourth* or *fifth* day there is more redness, and the vesicular appearance is more apparent; a round elevated ring forms with a depressed centre; on the *eighth* day the vesicle is ripe, it is more or less distended with lymph, and is surrounded with a red areola. Mr. Hands tells us that Jenner "was fond of likening the true cow-pox vesicle to a pure pearl placed upon a healthy blooming rose leaf." By the ninth or tenth day the disease has reached its height; there is often considerable redness and swelling, at times extending almost from shoulder to elbow. This subsides gradually but often quickly. The redness varies from an erythematous blush to a deep erysipelatous-looking crimson redness. The lymph becomes more opaque and gradually dries up, a scab forming about the *fourteenth* or *fifteenth* day, which falls off in eight or nine days leaving a red mark, which gradually becomes a whitened cicatrix, marked over with whitish scars or pits in the whitened ground. A good mark should not be less than one third of an inch in diameter.

The areola is sometimes the seat of a vesicular eruption j this was very well marked in the child of the lady above referred to.

In some cases there is a roseolous rash over the body.

Tests of Successful Vaccination. — Great stress is laid on the appearance of the cicatrix, by authorities on the subject. It is asserted, and I am not prepared to dispute the statement, that those with large and good marks escape better in attacks of small-pox than those with few and imperfect ones, and we are told not to be satisfied with one vesicle. I do not know what the practice of these gentlemen is: if they make five punctures, and but one happens to take, the child is not seen for a week, do they then re-vaccinate the child? We know that at a certain stage a second vaccination will overtake the first; if they are too late to do this it will be no use revaccinating, inasmuch as the whole system having been affected, though this is only shown by one small vesicle, no fresh vaccination will take till the effect of the first begins to wear out It is pretty clear that with one vesicle the child will be protected for a certain number of years, but that at the expiration of that time the necessity for re-vaccination will be more urgent than in cases where five vesicles have been produced, if we admit that it is proved that the greater the intensity with which vaccination takes, the greater the security.

Revaccination. — The necessity for revaccination is at once apparent from the number of persons who take small-pox who have been vaccinated. This number must be very large when it is stated that 50,000 vaccinated persons have died of small-pox within the last few years. — *Letter of Dr. Wyld in Daily News,* April 6, 1877.

It has been known for a long time that the protection obtained from vaccination is apt to die out, but of late years the evidence is becoming more conclusive, so that re-vaccination is more necessary now than it was some years ago. One fact, I think, proves this satisfactorily. About twenty-five years ago, the late Dr. George Gregory and Dr. Copland maintained that small-pox was never seen under puberty in the case of those who had been vaccinated. This rule no longer holds good, many children who have been vaccinated are now found to suffer from small-pox. This shows a great change in the power of the vaccine, and I imagine is due to the lymph wearing out.

Daring the last epidemic we used to hear it stated that a second well-performed vaccination was sufficient to protect during life; we never hear this now.

We also heard, a few years ago, that vaccination had been so well carried out in Ireland that there was no outbreak of small-pox when we were suffering in England. A very short time was sufficient to dissipate this cherished delusion. When the epidemic wave reached Ireland vaccination proved to be but a partial protection.

In the return form of births and deaths in the metropolis for the preceding week, March 7, 1877, 84 deaths were attributed to small-pox; out of that 37 were certified as unvaccinated, 23 as vaccinated, and 24 were "not stated" as to vaccination.

The deaths of fifteen unvaccinated children under five years of age were referred to this disease.

From another weekly return in the early part of 1877 we may take the following: —

"The Registrar-General in his weekly return points out that the deaths from small-pox, which had been 97, 75, and 116 in the three preceding weeks, were 100 last week, of which 34 were certified as unvaccinated, 31 as vaccinated, and in the remaining 35 cases the medical certificates did not furnish any information as to vaccination. In 29 of the 52 fatal cases of small-pox occurring last week in private practice in London, no information as to vaccination was given in the medical certificates. In comparing, however, the numbers of deaths occurring among vaccinated and unvaccinated persons, it should be remembered that there are probably now in the London population fully nine times as many vaccinated as unvaccinated persons."

These are probably fair samples of the weekly returns.

Considering the difficulty of ascertaining whether vaccination had been performed (unless in the case of the children), as it would be difficult to detect a trace of one or two small vesicles, and as 24 in one case and 35 in the other were "not stated," if the real number of those who had been vaccinated could be known, there probably would be a considerable increase over the numbers 23 and 31 returned as vaccinated; but taking these even as correct, it is a very large number out of 84 and 100, and while pointing to the desirability of re-vaccination, shows something very defective in the vaccination we have of late years been depending on. The argument is not used as against vaccination, but rather to show that there is something in our present system which needs amendment.

To account for the deterioration of the lymph, we must remember that probably much of that now in circulation has never been renewed since Jenner's time. Thus if some of the original lymph were used from *"arm to arm"* as fast as it could be used, since Phipps was vaccinated from the hand of the milkmaid who took the disease from the cow, now over eighty years ago, it may have passed through 4,160 individuals.

However, even admitting that the lymph might and ought to be much more reliable than it is, there is abundant evidence to show that revaccination rarely fails to protect the recipient of it for a number of years. I have known but one case where smallpox was said to have occurred within three or four months after revaccination. I believe other cases are on record, but they are so exceedingly rare that they are no argument against it.

Individuals vary wondrously as to their receptivity of second vaccination. For some five or six years, or even more after the first, we cannot expect a second to succeed; but after a longer interval we find some who will have arms like infants, others will have an areola with an acuminated spot or an abortive vesicle; with this there may be much itching, redness, tumefaction, pain down arm, and swelling in axilla.

In some the vesicles appear early and imperfectly, and finish their course two or three days earlier than in the infant.

It is much more common to find bad arms after a second than after a first vaccination; this may be partly owing to improper use of the arm, but it is so common that it is always well to give warning beforehand that a little extra pain or soreness must not cause surprise.

Injurious Effects of Vaccination. — Much is laid to the charge of vaccination by its opponents that loses weight by the manifest bias of the writer, and the suspicion of very gross exaggeration that attaches to what is said; on the other hand, its supporters do not always handle the subject as dispassionately as they might, and it is very evident that prejudice and partisanship have been allowed to influence the judgment, where above all things it is most important that we should get the real truth. How wisely Dr. Martin, the editor of Latham's works, says, "From controversy often comes exaggeration. And exaggeration often does the work of falsehood unawares! (Editor's preface, ix. *New Sydenham Society's Edition.*)

That syphilis may be conveyed by vaccination I think there is no doubt, though it must be of the rarest possible occurrence, for the reason that lymph would not be taken from a child presenting such unequivocal symptoms of disease as are likely to show themselves. In some cases these symptoms may be dormant; when this is the case, how far the lymph would be injuriously active it is difficult to say. Some say that if the lymph only is used without any blood there is no risk; this is simply conjecture.

Eczema and Porrigo may appear after vaccination. It would be very difficult to prove that they have been conveyed by it; on the other hand, we cannot deny the possibility; but having had the opportunity of making inquiries in several cases that came under my observation while in charge of the children's department at the London Homoeopathic Hospital, where various forms of skin disease were attributed to vaccination, I was often able to find evidence of the existence of the disease before the child had been vaccinated, and to show the mother that the vaccination had but called into greater activity, for the time, what was already there. I am not prepared to say that this happens always, for of course eruptions will follow vaccination, that we cannot show existed before, and that it is difficult not to attribute to vaccination. In the same way in a few cases *dregs* or sequelae of vaccination may show themselves, springing up in consequence of some constitutional delicacy.

Erysipelas, unhappily, follows vaccination on some rare occasions, and death, as we are aware, has taken place from it. In a recent case, it was shown that the medical man who had performed the operation prepared his points for using a second time, and after vaccinating put some of the bloody points when done with into the bottle with his charged ones. The wonder is that erysipelas is not more common, when we know what trifling causes will sometimes produce it. And as a large amount of inflammation is excited by vaccination, were the erysipelatous redness that we often see due to any other cause than its running a regular course of development and subsidence, we might expect to see more serious results.

It is, of course, possible that a cause of this kind, or an insufficiently cleaned lancet might be at fault; but that erysipelas has resulted from vaccination cannot be denied.

In the case of the late Sir Culling Eardley death was caused by pyaemia. My friend Dr. W. Bell, of Eastbourne, saw a somewhat similar case where there was profuse discharge, that happily recovered.

A patient of my own suffered severely from boils and thecal abscess after revaccination. Sending two of his sons to me to be vaccinated he expressed himself satisfied that in his own case he had had quite enough of it, and did not want to be done again. I have seen some very bad arms after revaccination; in one case an eruption like syphilitic rupia appeared: this was evidently called into activity by the vaccination, but not introduced into the system by it.

A gentleman who I had before attended in acute eczema wished to be vaccinated; I told him there was a possibility of a return of the eczema; he, however, decided to be done; the vaccination was followed by the most severe attack of eczema I ever witnessed; his sufferings were very great, and of many months duration. All medical men who have had much to do with revaccination must have seen a good many bad arms.

In the infant we may have a good deal of acute inflammation, but this soon subsides, and the arm does not continue sore for so long a time as we see in the adult. I shall have to say a little more on this subject when speaking of compulsory vaccination.

Vaccination as a Protection immediately after Exposure to Infection of Small-pox. — I recently vaccinated two elderly ladies the second day after they had unknowingly passed an evening in company with a small-pox patient on whom the eruption was coming out; the vaccination took well in both cases, and there was no small-pox. In the fifth volume of the *Annals of the British Homoeopathic Society* there is a paper by Dr. Harmar Smith, of Margate, where the desirability of vaccinating a patient sickening for small-pox is discussed.

Where this has been done the two diseases have gone on together, and the late Dr. Gregory inoculated one patient and vaccinated another from the eruptions on the one individual I believe the same thing was done by Dr. Smith. From what this gentleman says there is reason to think that it is well worth while to vaccinate in the early stage of small-pox, and as a test of what would be homoeopathic practice I should be glad to see the experiment tried on a large scale in one of our hospitals.

Our Supply of Vaccine Lymph. — At present much of our supply, as we have seen, is received from a source that is deteriorating.

The public mind is also far from being at rest as to the purity of the lymph used. Medical men are pretty often reminded of this by the caution given by mothers, "Be sure you get the matter you do my child with from a healthy baby." In the hurry and scramble to get a supply there is at least a temptation to take what lymph can be got; and though we trust no medical man would willingly take it from an unhealthy child, we cannot feel so sure as to the care he would employ, or the opportunities he would have for exercising it. When a strong healthy-looking baby is presented for vaccination it is not easy to get evidence as to the health of the parents. If we ask are both parents healthy, we may get an answer in the affirmative that is perhaps wide of the truth though answered in all good conscience, or the truth may be knowingly kept back. We cannot question a mother, who is perhaps the only one we see, as to her husband's history before marriage. All that can be said is that with so much doubt and uncertainty we may be very thankful there is so little known mischief; and each medical man must honestly do the best he can to obtain a pure supply.

To remedy the evil we naturally look about to see what can be done to obtain a more unfailing and purer supply. Just now we hear much of the Belgian lymph that has suddenly become famous.

The institution in Brussels for supplying it is under the charge of Dr. Worlomont, to whose courtesy in furnishing information different medical men in this country are indebted. The establishment was opened in 1868, but the vaccination has been kept up from calf to calf since the matter was obtained from a cow found to be affected with the disease in 1866. Medical men are supplied with lymph from the calves, and hitherto the vaccination has been carried on with very satisfactory results, though there is reason to believe that some of the difficulties we have experienced here were encountered there before final success was attained.

After various efforts to vaccinate the calf with the Belgian lymph in this country had failed. Dr. Wyld, who has taken much trouble in this matter, informs us that several medical men have succeeded in vaccinating calves. He also informs us that Dr. Martin, of Boston, U.S. (the same gentleman, presumably, alluded to earlier), has for seven years used lymph from heifers with good results, and he hopes that a supply from this source may be kept up.

The attempt to vaccinate the human subject with this lymph has also failed in the hands of expert vaccinators. This may be owing to the supply being received in tubes, in which way preserved, animal lymph appears to be less efficacious than when taken on points.

There can be no doubt that the obtaining pure lymph in this way would be an immense advantage. Hitherto this lymph appears to have given security in Belgium, but whether it will stand the test of frequent transmission from calf to calf has yet to be proved; it may be twenty years or more before this can be tested, but as the lymph originally taken from the cow and passed through the human subject is less powerful than it was, the same may happen with the calf lymph.

If, as I believe, the original disease in the cow comes from small-pox in the human subject, then there is the same possibility of its wearing out as there is of human vaccine; therefore, to guard against this, the proper course is to go back to the cow whenever opportunity offers, or to induce the disease by inoculating the cow, and then pass it through the calf, so that by a second remove from the human subject it would become milder.

In the event of a supply being needed during the prevalence of a variolous epidemic amongst cattle, inoculation of the cow should be performed in some part of the country where the disease has not yet appeared, and cows so operated on should be kept isolated till all fear of their spreading infection to other cattle would be over. The propagation of the disease by vaccination from cow to cow might be resorted to, to check its spreading in so severe a form amongst the animals themselves, as when taken naturally.

Too much importance cannot he attached to this question. For if the most efficacious supply, and that nearest to regular inoculation, is to be obtained in this way, then we can always obtain a supply by inoculating the animal, and by the steady use of this improved mode of vaccination small-pox may become a very rare disease.

Failing the carrying out of some efficient plan to supply pure and reliable lymph that would have a better repute than "Privy Council lymph" — which, however, with all its defects, has done good service [4] — we must be prepared to hear a good deal said against vaccination, and possibly a demand for some alteration in "the Vaccination Act of 1841."

Compulsory Vaccination. — Vaccination was made compulsory in 1833. Since then the regulations have been made more stringent, and have been carried out with considerable rigour.

There is no doubt that much unfair and mischievous opposition has been raised against it, and the prejudices of ignorant people have been roused instead of being quieted. On the other hand, the advocates of vaccination, having the law on their side, have carried matters with a high hand, no allowance is made for what people believe, and think they have good evidence for believing, and just cause of complaint is ignored, so that it is not to be wondered at that the anti-vaccinators secure a sympathetic audience amongst those on whom the act presses most heavily, and who are the ones least likely to understand its advantages. That the complaints are not without foundation we can see by enumerating them.

Thus, as a large number of vaccinated persons still die from small-pox, its efficacy is open to question.

That syphilis has been communicated by vaccination is admitted.

That at times skin diseases follow vaccination is believed.

That erysipelas, ending in death, has, on different occasions, resulted from the operation is true.

Also there may be a general disturbance of a child's health for some time, causing anxiety.

In addition to these not unreasonable causes of opposition to vaccination, there are others that will agitate the mind, and make some hesitate as to whether it is not better to pay a fine than allow their child's health to be, as they fear, destroyed.

Thus the dread of scrofula being engendered, or a constitution ruined by the vaccine taken from an unhealthy child, forms indefinite sources of trouble; and when there is apparently a real demonstration of the mischief that is said to arise by the exhibition of a sickly infant, or one covered with some horrid-looking eruption such as porrigo may show — the mothers of these children, truthfully as they believe, assuring a wondering and excited group of out-patients, in the waiting-room of a dispensary or hospital, that all that the eye sees is due to vaccination — is no allowance to be made for the feelings thus aroused? and is no effort to be made to correct this legitimate belief as to the reality of the danger?

I believe a little considerate thought for what others may feel, coupled with some judicious but not forced explanation, will go a long way to remove the apprehensions that exist in the minds of people; at least, I have found it so, and have succeeded in convincing some of the advantages of vaccination that I should have failed to make any impression on by simply opposing their opinion.

When I see the want of sympathy with the uneducated opponents of vaccination, my only wonder is that the number of antagonists to it is not largely increased.

The fact is that the dread of small-pox is such a powerful ally to the supporters of vaccination, that the great mass of the people will risk all the real and imaginary evils that the anti-vaccinators place before them, and will have themselves and their children vaccinated rather than incur what they justly believe to be the greater danger of remaining without the proved protection it affords. Were it not for this, compulsory vaccination would be doomed*

I do not like the Act in its present shape, and would be glad to see some modification in it. Thus, one fine ought to be sufficient, and a discretionary power might be left in the hands of the magistrates as to its infliction. I would vastly prefer persuasion to force in a matter of this kind.

The Registrar-General very fairly presses the point that out of the comparatively small number of those unvaccinated a large number die, while out of the enormous number of those vaccinated a much smaller number die; showing how terrible the disease would be but for vaccination. A fact of this sort brought well before the people is better than a hundred fines rigidly enforced. [5]

I would as a matter of wisdom, and as a matter of justice to the people, do all that can be done to make vaccination popular and safe, and by lectures and short explanatory pamphlets endeavour to remove prejudice and place the truth fairly before the people, doing everything to guard against the evils that have been enumerated, and would establish depots for the supply of pure vaccine from the cow or calf and from children, and would even go so far as to pay the mothers of healthy babies for the privilege of taking a supply from their arms, thus putting a value on good lymph, in order that the people might learn to appreciate it. I would then close every appointment or situation that possibly could against unvaccinated persons.

In this way I think prejudice against the system, as well as justly grounded complaints, would be removed, individual freedom would not be trampled on, and we might hope to see the terrible and loathsome disease of small-pox kept under control by the means mercifully placed within our reach.

[1] The same fact is no doubt told in Barron's life of Jenner, bat an additional interest is lent to Mr. Hands' statement, as he tells ns he subsequently vaccinated the children of this same Phipps.
[2] This is rather dangerous ground for experiment, but some interesting investigations have been made, hitherto without much result in the human subject. Mr. Hands tells us that distemper in dogs is arrested — indeed, prevented, by vaccinating the animal on the inner side of the thigh, and quotes M. Felizet for the fact that beasts affected with cow-pox are not attacked with foot and mouth disease. Thirty beasts were vaccinated, twenty-five successfully; not one of these, though exposed to contagion, was affected. How far there may be any identity of origin we can only suspect; further investigation is needed. Dr. Murchison thinks there is a connection between small-pox or rinderpest.
[3] This had been done by others previously.
[4] It is only fair to say that, as far as microscopic examination goes, this lymph is subjected to such a test, so that there is at least security that lymph is supplied, though the purity of its source cannot be thus Touched for. I have never had any reason to doubt its efficacy, nor have heard this questioned. The objection to it is the doubtful nature of the channel through which the supply passes.
[5] Though I have expressed a doubt earlier as to the exact accuracy of these reports, in the main they are sufficiently near the mark to fully justify the Registrar-General's conclusions.

Lightning Source UK Ltd.
Milton Keynes UK
UKHW031951241022
411033UK00001B/23